FOREWORD

In 1985, the OECD Nuclear Energy Agency (NEA) sponsored a first seminar on interface questions in nuclear health and safety, which provided an opportunity for experts in the fields of radiation protection, nuclear safety, and radioactive waste management to exchange views on several issues of common concern. These discussions have contributed to an improved dialogue between the different disciplines, and have helped the experts to work more effectively towards consistent positions on a number of questions.

In recent years, however, the regulatory implications of certain developments in radiation protection have increased uncertainty and caused concern throughout the nuclear energy field. The concerns centre on the current deliberations within the International Commission on Radiological Protection (ICRP) with regard to the revision of its basic radiation protection recommendations and the re-examination of the radiation risk factors based on data from revised dosimetry measurements on Hiroshima/Nagasaki populations. Other issues include the possibility of expanding the system of radiation protection to cover potential exposures, the definition of dose and radioactivity levels that are below regulatory concern, and the growing attention to radon exposure in buildings.

At the same time, there have been several developments in nuclear safety, both to prevent and to manage severe accidents that might occur, and to mitigate their consequences. In addition, efforts are under way to define quantitative safety objectives that are directly related to and are affected by the deliberations in the field of radiation protection.

The NEA therefore found it appropriate to organise a second interface seminar to improve the mutual understanding of the meaning and implications of current developments in radiation protection and nuclear safety and to build a framework for consensus between the two disciplines.

These proceedings include the papers presented and provide a record of the discussions held during the seminar. They are published under the responsibility of the Secretary-General.

THE INTERFACE IN NUCLEAR SAFETY AND PUBLIC HEALTH

◆

L'INTERFACE ENTRE LA SÛRETÉ NUCLÉAIRE ET LA SANTÉ PUBLIQUE

PROCEEDINGS OF THE SECOND NEA SEMINAR
COMPTE RENDU DU DEUXIÈME COLLOQUE DE L'AEN

PARIS, FRANCE
12th-13th September 1990

12-13 septembre 1990

NUCLEAR ENERGY AGENCY
ORGANISATION FOR ECONOMIC CO-OPERATION AND DEVELOPMENT

AGENCE POUR L'ÉNERGIE NUCLÉAIRE
ORGANISATION DE COOPÉRATION ET DE DÉVELOPPEMENT ÉCONOMIQUES

Pursuant to Article 1 of the Convention signed in Paris on 14th December 1960, and which came into force on 30th September 1961, the Organisation for Economic Co-operation and Development (OECD) shall promote policies designed:

— to achieve the highest sustainable economic growth and employment and a rising standard of living in Member countries, while maintaining financial stability, and thus to contribute to the development of the world economy;

— to contribute to sound economic expansion in Member as well as non-member countries in the process of economic development; and

— to contribute to the expansion of world trade on a multilateral, non-discriminatory basis in accordance with international obligations.

The original Member countries of the OECD are Austria, Belgium, Canada, Denmark, France, Germany, Greece, Iceland, Ireland, Italy, Luxembourg, the Netherlands, Norway, Portugal, Spain, Sweden, Switzerland, Turkey, the United Kingdom and the United States. The following countries became Members subsequently through accession at the dates indicated hereafter: Japan (28th April 1964), Finland (28th January 1969), Australia (7th June 1971) and New Zealand (29th May 1973). The Commission of the European Communities takes part in the work of the OECD (Article 13 of the OECD Convention). Yugoslavia takes part in some of the work of the OECD (agreement of 28th October 1961).

The OECD Nuclear Energy Agency (NEA) was established on 1st February 1958 under the name of the OEEC European Nuclear Energy Agency. It received its present designation on 20th April 1972, when Japan became its first non-European full Member. NEA membership today consists of all European Member countries of OECD as well as Australia, Canada, Japan and the United States. The Commission of the European Communities takes part in the work of the Agency.

The primary objective of NEA is to promote co-operation among the governments of its participating countries in furthering the development of nuclear power as a safe, environmentally acceptable and economic energy source.

This is achieved by:

– encouraging harmonisation of national regulatory policies and practices, with particular reference to the safety of nuclear installations, protection of man against ionising radiation and preservation of the environment, radioactive waste management, and nuclear third party liability and insurance;

– assessing the contribution of nuclear power to the overall energy supply by keeping under review the technical and economic aspects of nuclear power growth and forecasting demand and supply for the different phases of the nuclear fuel cycle;

– developing exchanges of scientific and technical information particularly through participation in common services;

– setting up international research and development programmes and joint undertakings.

In these and related tasks, NEA works in close collaboration with the International Atomic Energy Agency in Vienna, with which it has concluded a Co-operation Agreement, as well as with other international organisations in the nuclear field.

TABLE OF CONTENTS

TABLE DES MATIERES

Session 1 – Séance 1

THE INTERFACE ISSUE

———

LA QUESTION DE L'INTERFACE

Chairman – Président : F. COGNÉ (France)

5

THE PROBLEMS OF ACHIEVING NUCLEAR SAFETY OBJECTIVES

Session 2 - Séance 2

BASIC OBJECTIVES AND POLICIES

OBJECTIFS ET POLITIQUES : ASPECTS FONDAMENTAUX

Chairman - Président : R.H. CLARKE (United Kingdom)

THE ICRP RADIATION PROTECTION PHILOSOPHY AND ITS APPLICATIONS

THE UNITED STATES SAFETY GOALS

EUROPEAN SAFETY APPROACHES

THE PRINCIPAL NUCLEAR SAFETY POLICIES IN JAPAN

Session 3 – Séance 3

ACHIEVEMENT OF RADIATION PROTECTION
AND NUCLEAR SAFETY OBJECTIVES
IN THE REGULATORY AND TECHNICAL PRACTICE

REALISATION DES OBJECTIFS DE LA RADIOPROTECTION
ET DE LA SURETE NUCLEAIRE DANS LA PRATIQUE
AU PLAN REGLEMENTAIRE ET TECHNIQUE

Chairman – Président : D. BENINSON (ICRP)

7

Session 4 – Séance 4

ACHIEVEMENT OF RADIATION PROTECTION
AND NUCLEAR SAFETY OBJECTIVES
IN THE REGULATORY AND TECHNICAL PRACTICE (Contd.)

———

REALISATION DES OBJECTIFS DE LA RADIOPROTECTION
ET DE LA SURETE NUCLEAIRE DANS LA PRATIQUE
AU PLAN REGLEMENTAIRE ET TECHNIQUE (Suite)

Chairman – Président : H. KOUTS (United States)

Session 5 – Séance 5

CONCLUSIONS AND RECOMMENDATIONS

———

CONCLUSIONS ET RECOMMANDATIONS

Chairman – Président : K.B. STADIE (NEA)

OPENING SESSION

SEANCE D'OUVERTURE

OPENING REMARKS

K. Uematsu
Director General
OECD Nuclear Energy Agency

Ladies and gentlemen, it gives me great pleasure to welcome you to the Second NEA Seminar on Interface Questions in Nuclear Health and Safety.

After a first meeting in 1985, we are making this second attempt to assist experts in radiation protection and nuclear safety in jointly reviewing issues which are of common concern to both disciplines. I am happy to note that the earlier meeting contributed to an improved dialogue between both sides and has - most important for our Agency - stimulated closer co-operation between our competent committees in these areas: the Committee on Radiation Protection and Public Health (CRPPH) and the Committee on the Safety of Nuclear Installations (CSNI), and I am sure that this co-operation will extend to the Committee on Nuclear Regulatory Activities (CNRA), which was provisionally set up by our Steering Committee for Nuclear Energy.

Recent years have seen a number of changes in radiation protection and nuclear safety which make our encounter, today and tomorrow, even more timely. Developments in radiation protection have, notably, increased uncertainty and caused concern in the nuclear field and these should be aired openly. First and foremost among these developments is a revision of the basic radiation protection recommendations by the International Commission on Radiological Protection (ICRP) which are largely based on the recent re-examination of the radiation risk factors resulting from revised Hiroshima/Nagasaki dosimetry and expanded epidemiological information. There are also other questions arising from the radiation protection community, such as the possibility of extending the system of radiation protection to cover probabilistic exposures, the definition of dose and activity levels that are below regulatory concern, the growing attention to radon exposure in buildings and the transition from the intervention criteria used in emergency situations to the radiation protection regime in the far field and in the long term, in the aftermath of a nuclear accident.

Equally, there have been new developments in nuclear safety regarding the prevention of accidents and, even more important, regarding the management of severe accidents. Probabilistic safety assessments, as well as TMI and Chernobyl, have taught us that we cannot totally rule out such extreme events and much effort has, therefore, been devoted in recent years to the mitigation of the consequences of severe accidents, both through measures on- and off-site. Several countries are also in the process of adopting, or have adopted, quantitative safety objectives which are closely related to decisions in the field of radiation protection.

You will have no shortage of questions to consider over the next two days and I will not take up any more of your time, but please be assured that we, in the Agency, stand ready to assist in any way possible in enhancing future co-operation. Let me close by wishing you an informative and inspiring dialogue. I look forward to seeing you all tonight at the reception which I am happy to host in Room A of the Chateau.

SETTING THE SCENE

K.B. Stadie
OECD Nuclear Energy Agency

ABSTRACT

Given that the risk of a nuclear plant having an accident and of an individual inducing cancer from a small radiation dose are essentially probabilistic in nature, the paper advocates to define a probability threshold for the assessment of harm for the following reasons: 10^{-6} to 10^{-7} per year represents the lowest level of risk at which major engineering projects can be assessed with our present knowledge. At the same time, it is at least uncertain whether a large number of very small exposures really result in a collective detriment which a linear hypothesis would suggest. It would therefore be more correct to recognise these limitations and treat these low probabilities differently from the range which is better understood. Such a distinction – if agreed – is also likely to enhance the nuclear debate which is presently suffering from the open-ended risk assumptions pursued by the nuclear industry.

Among the many issues between radiation protection and nuclear safety, the question of how to limit risk considerations seems particularly important.

EXPOSE DU CONTEXTE

RESUME

Etant donné que le risque qu'une centrale nucléaire ait un accident et qu'un individu contracte un cancer à cause d'une petite dose de rayonnement est essentiellement de nature probabiliste, la communication préconise de définir un seuil ·de probabilité pour l'évaluation des dommages pour les raisons suivantes : 10^{-6} à 10^{-7} par an représente le plus petit niveau de risque auquel les projets importants d'ingénierie peuvent être évalués sur la base de nos connaissances actuelles. En même temps, il est pour le moins incertain si un grand nombre d'expositions très petites aboutissent réellement au détriment collectif suggéré par une hypothèse linéaire. Il serait par conséquent plus correct de reconnaître ces limitations et de traiter ces petites probabilités de manière différente que le domaine qui est mieux compris. Une telle distinction – si elle est acceptée – est également susceptible d'améliorer le débat nucléaire qui souffre actuellement des hypothèses de risque sans limites examinées par l'industrie nucléaire.

Parmi les multiples points de débat entre la radioprotection et la sûreté nucléaire, la question de la limitation des considérations de risque semble particulièrement importante.

Radiation protection and nuclear safety represent the two sides of the same coin. They are complementary in the protection of man and his environment, and as such, they together can make the exploitation of nuclear fission acceptable from the risk point of view.

Both disciplines share a common objective, but they pursue it using different philosophies of risk management, and these differences, which I believe are more apparent than real, may be explained by historic developments, cultural differences, and by their different origins.

To begin with, we must recall that the sophisticated safety techniques used by the nuclear power industry are the end results of a long series of efforts begun by the early inventors of machines and structures, to prevent accidents. These pioneers were often carried away by their zest, harming first of all themselves and then others. Of course they were handicapped by their limited knowledge and as a result, the road to progress in many technologies was paved by numerous accidents and disasters which in turn provided the necessary lessons to do better next time. There is, perhaps, a legacy here for which we still pay.

Overdesign (the provision of design margins) and diligence at all stages – natural for the pioneers – were the first means by which they tried to prevent failures. But, as structures, and particularly machines, increased their scope and size, potentially endangering a larger and larger number of people – including those not directly involved – added margins and diligence alone did not suffice.

Ultimately, however, all industrial incidents and accidents can be traced back to human fallibility or the limits of human understanding.

A new approach was therefore needed to control technological risks by making the possibility of accidents very remote. The accidental behaviour of machines was increasingly analysed and a myriad of systems, components and barriers developed and added, to circumvent human shortcomings and lack of understanding, thus reducing the probability of harm. Diverse and redundant means of shutting down machines were devised and harmful substances were increasingly separated from the biosphere by several independent barriers each protected by safety systems, assuring that they are not challenged beyond their limits. This defense-in-depth philosophy constitutes an integral part of today's industrial development, with the result that modern technology has attained an extremely high resistance to catastrophic failures.

The nuclear industry has built on this foundation and has raised the concept of safety technology to a fine art, both for accident analysis and for ways to limit the probability of accidents and mitigate their consequences. Even beyond these precautions, the siting of nuclear power plants away from population centres was accepted as an additional means of limiting the risk from the beginning.

The early efforts in nuclear safety were aimed at defining the scope of safety considerations, as had been the practice for all previous technologies. This led to the definition of the design basis accident, which was described as the worst credible accident which might occur to a reactor during its operating life. The DBA concept served primarily two purposes: to focus safety-thinking and to direct and limit the safety design.

Inherent within the DBA concept is the requirement that the installation must be able to withstand the design basis accident without causing major harm to the public. In turn, anything worse than the DBA was considered to be so unlikely that it was classed incredible, which in turn meant that there was no need to take specific precautions against it or its consequences.

More recently, the emergence of probabilistic risk assessment, and the TMI accident, have gradually broadened the scope of safety considerations. Implicit in PRA is the acknowledgment that the mere fact of using nuclear energy implies the acceptance of some finite degree of risk and that therefore there is no logical way of differentiating between credible and incredible accidents.

These developments have given impetus to the search for a safety concept which must obey two somewhat contradictory requirements. It must, on the one hand, consider how to limit the consequences of very rare severe accidents, larger than a DBA, and on the other, it must clearly distinguish these considerations from well-defined regulatory measures which necessarily address specific and well-defined accidents (DBAs), having probabilities closer to the realm of human experience.

This somewhat open-ended severe accident concept has - in recent years - provided an opportunity for society to request protection against events that have ever decreasing probability.

Reg Farmer, one of the key architects of PRA, has insisted, for the past few years now that the probability range of 10^{-6} and 10^{-7} per year represents the lowest level of risk at which major engineering projects can be assessed. We do not know enough about equipment and system behaviour, let alone human behaviour, to do better and he thus has grave reservations about any further extension, given the state of knowledge. An increasing number of safety experts therefore argue that the 10^{-7} or 10^{-8} probability level could provide a level below which a change of treatment for nuclear safety considerations is required because it becomes increasingly difficult to identify possible initiators at or below 10^{-6} per year and because it becomes difficult or impossible to assess consequences of certain very severe events of extremely low probability.

In contrast, the radiation protection field derives its origin from medicine, which has always occupied a special place in society. Its lifesaving qualities have been perceived long before Aesculap and medical doctors, together with a limited number of reseachers in physics, at the turn of the century, were the first to knowingly suffer from man-made radiation. They were the first to react and to set limits for their use - more than 80 years ago nationally and more than 60 years internationally. Or, in other words, efforts were well underway to limit the radiation risks before fission was known to be capable of producing energy on a large scale.

Early radiation protection effects were concerned with high doses which were proven to cause biological harm. Time of exposure, shielding and distance from the source were the means to limit the risk. With the growth of epidemiological studies and spurred on by increasing use of fission energy, quantitative approaches were introduced in the 1970s.

Nature is by far the largest radiation source. Background radiation differs more than an order of magnitude between different regions of the globe and also varies with time. Despite this vast statistical basis, research has not been able to provide conclusive proof that the radiation risk is proportional to exposure. However, because one can rationally accept a probabilistic model for the induction of cancer at the level of a single cell, the ICRP has prudently promulgated a radiation protection philosophy which is based on the hypothesis of proportional correlation between risks and exposure at low radiation levels.

Radiation protection and nuclear safety have thus evolved along analogous lines. Both began by setting limits to radiation exposure or safety challenges and both have gradually broadened the scope by recognizing that it is not intellectually honest to arbitrarily define a threshold; because just as any dose must be presumed to result in some probability of fatal cancer, the operation of a nuclear plant must be presumed to involve some probability of an accident.

Given that the risk of a nuclear power plant having an accident, and of an individual inducing cancer from radiation are essentially probabilistic in nature, both disciplines now face the same challenge: how to define a level below which the risk cannot be defined well enough and thus requires to be treated differently.

We have seen that some in the nuclear safety community are trying to define a probability range below which the risk cannot be reasonably assessed. Is not the same true for radiation protection and isn't there a dose below which the induction of cancer is so small - if it exists at all - that it should be treated differently from the range that we understand?

Exemption levels, considered to be below regulatory concern, have recently been developed. They have, however, only an administrative meaning and a large sector of the radiation protection community does not see them as a threshold below which large numbers of very small individual exposures could be treated differently for the calculation of collective detriment. Yet, it is the absence of such a distinction that so often leads to misinterpretation and misuse.

It is evident that the approach by both the radiation protection and the nuclear safety community to assess and to cope with the radiation risks in the nuclear industry is uncommonly honest. We must therefore avoid that existing uncertainties are exploited by those who unrealistically and unnecessarily demand zero risk from our industry. To limit our range of action along the lines I have described could be of help here.

Ladies and gentlemen, you will debate many questions, which unite and which divide both disciplines, over the next two days. The Director General has alluded to them in his opening remarks. I have addressed but one aspect which I nevertheless consider to be particularly important.

Let me close with one observation which may be relevant to our debate. Radiation strikes humans individually and invisibly, while accidents are very visible since they often affect many people in a single event. Is it therefore surprising that the radiation community, which in addition is associated with the medical profession, is held in higher regard by our compatriots than engineers and scientists, who are accountable for safety? How do we bridge this gap?

We have put together a programme which we hope provides the opportunity to examine the various issues between both disciplines and to air the differences with the view to contributing to a well founded and compatible radiation protection and nuclear safety philosophy.

Session 1

THE INTERFACE ISSUE

Séance 1

LA QUESTION DE L'INTERFACE

Chairman – Président

F. COGNÉ
(France)

GENERAL OVER-VIEW ON INTERFACE ISSUES
BETWEEN NUCLEAR SAFETY AND RADIATION PROTECTION

R.H. Clarke
National Radiological Protection Board
Chilton, Oxfordshire, United Kingdom

A. Birkhofer
Gesellschaft für Reaktorsicherheit Forschungsgelande
Graching, Federal Republic of Germany

ABSTRACT

The philosophies for risk management in nuclear safety and radiation protection are different. In this paper, the historical background that has led to this situation is reviewed to identify the points of interface. The paper presents the principles of protection that are emerging from the new recommendations of the International Commission on Radiological Protection and demonstrates that these are compatible with those used for nuclear safety analyses. The conclusion is that the two communities of protection and safety are working along convergent not divergent lines towards answering the question "How safe is safe enough".

APERCU GENERAL DES QUESTIONS D'INTERFACE
ENTRE LA SURETE NUCLEAIRE ET LA RADIOPROTECTION

RESUME

Les doctrines en matière de gestion des risques dans le domaine de la sûreté nucléaire et dans celui de la radioprotection sont différentes. Dans la présente communication, les auteurs passent en revue les circonstances histori-ques à l'origine de cette situation afin de cerner les points d'interface. Ils exposent les principes de protection qui se dégagent des nouvelles recommanda-tions de la Commission internationale de protection radiologique et démontrent qu'ils sont compatibles avec ceux utilisés pour les analyses de la sûreté nu-cléaire. Ils en concluent que dans les disciplines tant de la radioprotection que de la sûreté nucléaire, les travaux sont menés selon des orientations con-vergentes et non pas divergentes en vue de trouver une réponse à la question "jusqu'où doit-on perfectionner la sûreté ?"

HISTORICAL BACKGROUND

Within a year of the discovery of x-rays in 1895, papers appeared in the literature reporting that high doses of radiation could cause biological harm. For the period from the beginning of the century up until the 1950s, it was assumed that the only harmful effects of radiation were the non-stochastic effects which occurred above certain thresholds of dose. Radiation protection controls were based on avoiding non-stochastic effects, both in routine operations and in accidents.

With the development of nuclear power for commercial purposes, starting in the 1950s, the potential for accidental releases was clearly recognised and, because of the concern about harmful effects of high doses of radiation, reactors were sited in areas of low population density. The philosophy of nuclear safety at this time was to keep the chance of accidents remote and to keep the possible number of people exposed low. Radiation protection philosophy was to keep doses below thresholds.

Most nuclear plants were therefore designed with certain dose criteria having to be met for both routine operations and for accidents. By the 1970s radiation protection had moved to a more quantitative risk basis with the emergence of radiation risk factors derived from epidemiological studies, such as those on the Japanese A-bomb survivors. It became possible with this quantification of radiation risk to set dose limits for routine operations to correspond to levels of acceptable risk. At the same time that this was happening in radiation protection, probabilistic risk assessment (PRA) began to emerge in reactor safety assessments. Safety engineers were developing the techniques whereby a spectrum of accidents could be looked at in terms of probabilities, and they began to adopt the radiation risk factors to enable them to express the results of their PRAs in terms of risk to the most exposed individual.

In the 1980s, the radiation protection community began to ask itself why the risk estimates which were applied for the development of dose limits should not also be used to limit the overall risk from accidents. At the same time, the nuclear safety community was beginning to ask itself what it could do with the PRA technique it had developed. Could it be used for siting? Could it be used for emergency planning? But, most important, could it be used to say how safe is safe enough?

PRESENT SITUATION

So here finally radiation protection philosophy and nuclear safety have met head on. The reactor safety community had been developing techniques for risk assessment which were scenario based, ie depended on each fault sequence and could be summed to give an overall level of individual risk from a single source. The radiation protection community saw risk limits as being analogous to dose limits and therefore applied to the total risk from all sources of exposure - nuclear power plants, interlock systems on radiotherapy suites, etc.

There are problems in using individual risk in that it is not meaningful to use an expectation value for very low frequency events because, at very low frequency, the events by definition will almost never occur and therefore the average of a large number of such events, ie expectation value, can have no scientific meaning. How then are we to deal with adding contributions of risks from events which may have large doses but only a vanishingly small probability of occurrence? Here the same question is faced by both radiation protection and nuclear safety communities. Various options can be suggested, of which risk cut-offs for individuals are sometimes proposed.

The radiation protection community had also emphasised increasingly over the years that compliance with limits was a necessary but not a sufficient condition for protection. Optimisation, or As Low As Reasonably Achievable (ALARA), was seen as the more important method for ensuring exposures were not too high. In routine operations, optimisation of protection is straightforward. One can evaluate the collective dose, attribute cost to it, and balance the expenditure on plant improvements against reductions in the cost of collective dose. No simple and straightforward formal approach has been found to optimise nuclear safety. However, optimisation has been implicitly done by the continuous search for better designs with improved technical solutions which increase safety margins - so decreasing accident frequencies.

So here again is the interface between radiation protection and nuclear safety. Techniques have to be found by both groups which can lead to answers to the question, how safe is safe enough.

CONCEPTUAL FRAMEWORK

The International Commission on Radiological Protection has released, in draft form, its proposed recommendations. There is a determined attempt to bring together all aspects of radiation protection into a single conceptual framework. However, the Commission recognised and accepted the difficulties created by its use of the terms Planned, Potential and Pre-existing, and the confusion in the definition of the three principles of protection in each case.

The new text will deal with PRACTICES which add doses and risks to people and INTERVENTION which subtracts doses that would otherwise be received. The definition of Practices would include nuclear power stations which give rise to doses which are in addition to any others that already exist to the workforce at the station and to members of the public from effluent discharges. On the other hand, examples of Intervention would be remedial action to reduce high radon levels in homes, or planning evacuation to avoid doses in the event of an accident at a nuclear plant. For both situations, a generalised set of principles will be expressed in the following style.

(a) Do more good than harm in making   (JUSTIFICATION)
     decisions involving radiation

(b) Obtain the best levels of protection in   (OPTIMISATION)
     the circumstances

(c) Keep the likelihood of harm below   (LIMITATION)
     unacceptable levels.

No doubt the exact wording will be agonised over at the next meeting of the Main Commission in November, but the idea will be that the principles apply in all cases, although different numerical values will result for interventions compared with practices.

It has to be said immediately that methods, procedures and input data for decision making do not exist for all situations where the principles are to be applied. In fact, justification is an extrmely broad principle involving political and strategic inputs where nuclear safety or radiation protection considerations will be only one of many inputs, and then they may only have minor influences on the decision-making process.

In the rest of this paper, use of the two principles of optimisation and limitation, (b) and (c) above, will be applied to the control of radiation risks to workers and the public in normal operations and in accident situations.

APPLICATION OF THE PRINCIPLES

Realisation for Workers

With regard to the protection of workers, the principle of limitation is realised by individual dose limits. Optimisation is controlled and reliably assured by the radiation protection organisation in any particular facility. Important reductions of the occupational collective doses (up to a factor 10) have been achieved in the past by extensive use and careful evaluation of operational experience in order to improve plant design, technical equipment and operational procedures.

In nuclear power plants for instance, changes of materials and water chemistry as well as optimised procedures have led to a large decrease of the coolant activity, exposure times have been reduced by providing space for manual work, and automated examination techniques have been developed. Those and similar efforts have provided a fairly optimised level of protection.

Most routine operations are internally optimised with respect to occupational exposure. Thus, further improvements depend highly on the development of integrated strategies addressing a whole spectrum of operations and the aspects of various fields such as radiation protection, engineering, maintenance and administration. Conflicting requirements from different aspects might occur. Therefore more detailed information is required on, for example, correlations between specific jobs and the doses received and identification of operational factors giving rise to radiation fields. There must also be rapid information exchange between the different professional groups involved. For instance, a better co-ordination between technical requirements for vessel inspections (which are based on consideration of conventional risks) and radiation protection could provide better optimisation, ie a further reduction of personnel exposure, which cannot be achieved by radiation protection alone.

Effluents

In nuclear facilities a multi-barrier concept is applied in order to reliably separate radioactive materials from the environment. In most facilities, in particular in nuclear power plants, those procedures have been considerably strengthened over time and the effluents have been decreased far below the limits set by regulatory requirements. Thus, its contribution to both individual and collective dose has become almost negligible compared with natural and other artificial sources.

In the vicinity of German NPPs for instance, the maximum individual effective dose is currently below 10 μSv/a, and for most plants below 1 μSv/a. It can be claimed that the effluents of such plants are optimised with respect to individual and collective doses. Attempts to further reduce effluents have to be carefully checked with respect to a possible increase of other nuclear and non-nuclear risks.

Even at higher levels of effluent discharge, the optimisation of doses can be a very complex task. An example is the krypton retention in reprocessing plants. The project of a German reprocessing plant which was cancelled last year was supposed to release about 160,000 TBq of Kr-85 per year. Distributed in the atmosphere, the annual release would lead to a global collective effective dose commitment of about 35 manSv in 5 billion people, corresponding to 1000 manSv accumulated over 30 years of operation. At 3,000 US$ per manSv, the equivalent of that dose commitment is 3,000,000 US$. The costs of a krypton retention plant were projected to be several tens of millions of US$ for investment and more than 1 million US$ per year for operation.

Nevertheless, it was decided to pursue the project of krypton retention on the basis of consideration of maximum individual doses, confirming the experience that individual doses normally prevail even in optimisation. However, there were many other factors to be considered for decision making. Besides the individual and collective doses of workers for operating of the plant, non-nuclear mishaps such as explosions with a potential of severe off-site releases had to be taken into account. As the possible technologies were relatively new and not yet proven, experience with such risks was scarce and considerable uncertainties existed about the related basic data used for decision making. In view of such difficulties, it was decided to construct a demonstration plant before planning full scale krypton retention.

It is a characteristic feature of decision making that the underlying data are to some degree uncertain. In an integrated approach to risk, those uncertainties have to be assessed consistently for all concerned aspects. The use of the collective dose alone as a measure of harm implies the value judgement that a high individual risk for few people is equivalent to a small individual risk for many people, provided that the collective dose is the same. That judgement might be questioned, in particular if the accuracy of risk prediction is significantly different in both cases and if the individual risk is extremely low.

Protection with respect to accidents

Traditionally, nuclear safety is realised by good design and quality of technical equipment and by appropriate operation assuring technical safety at normal operation as well as prevention of abnormal events, incidents and accidents, which could jeopardise the integrity of the successive barriers separating the radioactive material from the environment. This procedure of multi-level prevention supplemented by mitigation is known a defence-in-depth.

The design of nuclear facilities ensures that consequences of design-basis accidents are below unacceptable levels. On the other hand, at least in the western world, defence-in-depth is being continuously improved using operationsl experience and safety research wherever this seems reasonable. Thus, possibilities of mishaps on the various lines of defence are reduced over time. Those improvements in accident prevention have made design basis accidents very unlikely and reduced the related risk to extremely low levels.

The risk of present nuclear facilities is therefore basically related to what is called "beyond design", ie scenarios which have been considered too unlikely for introduction of additional safeguards in the plant design. The frequency and the consequences of such scenarios are considered within probabilistic risk analyses providing an estimate of individual, population and environmental risks.

Regarding nuclear power plants, which among the nuclear facilities have the largest risk potential, a number of probabilistic risk analyses have been performed during the last two decades. It turned out that the frequencies of a core damage accident are mostly in the range between 10^{-5} to 10^{-4} per year of reactor operation. Safety targets, striving for core damage frequencies below 10^{-4}/yr for existing and below 10^{-5}/yr for new reactors are actually used by the nuclear safety community.

The core damage frequency is an upper bound of the individual risk of a member of the population to die from an accident in a given nuclear power plant. Therefore, existing plants comply or can be made to comply with an individual risk limit of 5.10^{-5}/yr as proposed by ICRP, provided the consequences of an accident are limited to the single plant, ie no superposition of risk from different plants.

It is reasonable to consider the technical potential to further decrease the core damage frequencies and thus the individual risks by improvements in accident prevention. Nevertheless, it is evident that this should not be the only way of optimising protection against nuclear accidents in view of the question "how safe is safe enough". There are basically two reasons for seeking an additional way.

1. To improve prevention of accidents in order to decrease the core damage frequency much below 10^{-5}/yr is increasingly difficult, because of lack of sufficient data from experience to perform probabilistic analyses. Multiple failures for instance, which are almost never observed in reality, have to be assessed theoretically, often taking a leading role for accident propagation. It is further increasingly difficult to assure the completeness of the considered accident spectrum. Thus the uncertainty of calculated core damage frequencies increases considerably at very low frequencies, indicating a need for additional and independent accident mitigation.

2. In a nuclear accident, a very large number of persons may be involved and socio-economic impacts may be drastic. It is therefore necessary to reduce the number of persons at risk and to exclude the possibility of a severe deterioration of the social infrastructure in the event of such an accident.

The most promising way to overcome those problems seems to be the introduction of an additional level of defence-in-depth, largely independent of the success of reducing the probability of the accidents. Such a procedure can optimise protection with respect to large scale population and environmental impacts from accidents.

How can this additional level of protection be realised in practice? It might be by improving the preparation and organisation of off-site emergency measures. However, off-site countermeasures would not really remove the socio-economic impact of a large nuclear accident. On the other hand, precautions at the source are more efficient in order to control the hazardous materials than actions which are effective in the environment. The discussion about the filtered venting concept in the accident management of German NPPs illustrates that. Should the decision about the initiation of venting in the event of a severe core damage accident take account of the wind direction in order to minimise public exposure by the radioactive releases? There is a strong argument against it since the essential benefit of venting is the protection of the containment against overpressure. Opening the vent valves too late in order to take benefit of a more favourable wind direction could lead to a catastrophic containment failure overriding all beneficial effects of wind direction.

These arguments strongly suggest that the best additional level of defence is realised by emergency measures inside the plant, ie by accident management, including the mitigation of severe core damage. This approach should strengthen the containment function and make it less dependent of the history and the loads of severe accident sequences. An intact containment can sufficiently contain the consequences of a core melt accident, as the experience of TMI has shown. Therefore, this strategy can best be represented by limiting the conditional probability of containment failure in the event of a core melt accident. Proceeding this way constitutes the best actually known procedure of optimising radiation protection with respect to severe nuclear accidents.

The question still remains, how safe is safe enough? The possibilities of predicting risks and the resources to eliminate them are limited. With state-of-the-art probabilistic methods, it is hard to assess events with frequencies below 10^{-7}/yr. Results are accompanied by large error bounds. Therefore, it seems not reasonable to consider individual risks below a range of 10^{-7}/yr to 10^{-6}/yr.

On the other hand, it is intended to bring the core damage frequencies to below 10^{-5}/yr by plant modification and by improved concepts. The conditional probability of containment failure must then be below a range of 10^{-2} to 10^{-1}. This is achievable and represents a balanced approach, taking into account the actual state of technology and analytical methods. Together with the target for core damage frequency, such a requirement would assure that a severe core damage accident is unlikely to happen in the world during the lifetime of actual plants and – should it happen – would with high confidence not lead to socio-economic disruptions and major health effects in the population. The long-term objective should be to make unnecessary the planning of interventions outside the plant.

EXEMPTION LEVELS

There is continuing interest in the specification of levels of dose or risk which are sufficiently low to be "beneath regulatory concern". It is first necessary to clarify that there is a difference between a dose cut-off in collective dose calculations and the exemption of a practice or source from regulatory control. It is the second of these that is addressed here.

Most authors seem to agree that there can be a level of individual risk which is so small as not to be of concern to the individual. This figure generally seems to be about 10^{-6} yr^{-1}, although in some circumstances it can be lower or higher. In terms of dose, using current ICRP proposals for risk estimates, this corresponds to about $20\mu Sv.yr^{-1}$.

However, ICRP philosophy over the last 20 years has evolved to ensure that not only is each individual sufficiently protected, but the population is also adequately protected. This is because if each source is regulated so that only the individual most exposed from that source is protected, doses could build up from many sources to give doses that are not negligible.

This topic was addressed at the last NEA Interface meeting in 1985 and NEA subsequently devised, together with IAEA in 1988, criteria for the exemption of sources from regulatory licensing. In essence, the NEA/IAEA work formalised the way in which optimisation plays an important role in the decision to exempt a source. The important feature was the recognition that the optimisation process should include the cost of regulatory resources to be balanced against the cost of the collective dose from normal operation of the practice.

The result was that the NEA/IAEA definition of an exempt practice was one in which no individual received a dose greater than of the order of 10 microsieverts per year AND that a continuing practice gave rise to no more than a commitment of about 1 man sievert per year of the practice. NEA gave three examples where exemptions could be applied to advantage; consumer products, low level radioactive wastes, and re-cycling or re-use of materials.

In general, it is probably true that the individual dose criterion will be the more restrictive, since the numbers of people involved in any exposures will be limited. If, for example, a practice gives the most exposed individual 10 μSv yr^{-1}, the average exposure is likely to be of the order of 0.1μSv y^{-1} so that 10 million people would have to be exposed before the collective criterion was more restrictive. Even then, however, optimisation might show the practice should be exempted.

In the field of nuclear safety, it has been said that assessments are often scenario based and that an individual risk figure will be determined, but there is no comparable "collective risk" criterion that can be calculated analogously to collective dose. Since optimisation cannot be undertaken easily for probabilistic events, the individual risk cut-off will be used at present to determine those events that can be dismissed as being too small to warrant commitment of regulatory resources.

By comparison with the radiation dose case, if the level of trivial risk to an individual is agreed to be 10^{-6} yr^{-1} and this is from ALL sources, it follows that an exemption level for a single accident sequence is probably about 10^{-7} yr^{-1}.

CONCLUSIONS

Many people see radiation protection and nuclear safety as being in conflict. This is the wrong conclusion. Safety and protection have developed in parallel, rather than series, and have recently met, somewhat to the surprise of all concerned.

Both in radiation protection and nuclear safety there is the assumption of the absence of a threshold in risk. Just as any small dose of radiation must be assumed to give rise to an associated chance of cancer, so any nuclear plant has a probability of an accident. In both cases all must be done to ensure that doses or risks are as low as reasonably achievable. In both cases the real problem is to decide what is to be considered reasonable - how safe is safe enough?

For routine operations, ICRP has recommended optimisation and has recommended methods of multi-attribute analysis to accomplish it. ICRP is now trying to extend the techniques to accident situations. This begins to impinge on nuclear safety which, as has been said, has developed a scenario based procedure, but involves not only the probability of the sequence occurring, but also the probability of harm resulting if the accident occurs.

However, successful optimisation of protection against accidents is not easily amenable to simple and straightforward methods such as cost-benefit analysis. One intrinsic difficulty is the absence of a reasonably accurate and meaningful measure of the collective harm from an accident. In order to be succesful, optimisation has therefore to be used in a rather broad sense. The usual approaches by the nuclear safety community have been to search continuously for better solutions and technical progress beyond the mere fulfilment of formal requirements.

These features make it more difficult to argue the "safe enough" case. Nevertheless, effective mitigation of off-site consequences by accident management, which is largely independent of accident frequency, seems an important step in that direction.

There is now an intellectual challenge in establishing integrated risk management. The communities of protection and safety need to solve the problems jointly because if they are divided and argue, the result will be even more loss of public confidence.

DISCUSSIONS

F. COGNE, France

Cette communication est ouverte pour discussion.

R.J. BERRY, United Kingdom

While agreeing with most of what Dr. Clarke has said, I must take issue with two of his definitions.

1) Early in the paper he stated that reduction in dose <u>limits</u> was the basis of dose reduction. In the nuclear industry, we would argue that the overriding injunction to keep all doses "as low as reasonably achievable", ALARA, (ALARP in the UK) is the driving force for dose reduction and that excessively restrictive annual occupational dose limits can actually <u>increase</u> both individual, lifetime and collective doses.

2) He suggested that discharges from nuclear facilities had been "optimized". We would answer that discharges have been reduced to levels as low as technically achievable (minimised), irrespective of cost. In asking "how safe is safe enough", optimisation <u>must</u> include some consideration of the cost for further reduction in both indiv<u>i</u>dual and collective doses.

R.H. CLARKE, United Kingdom

I do not think I said that reduction in dose limits was the main reason for bringing down occupational exposure, at least I did not mean that. I hope I said that dose limits were being reduced by ICRP in the context of the principle of limitation. I agree with you, the principle of optimisation is thought to have reduced exposures over the years, both individual and collective.

For the second point, I believe it is generally true that for environmental releases, individual dose has been more important than collective dose in driving the reduction. Money has been spent to reduce critical group exposures: generally collective dose considerations has been less important, except perhaps for carbon-14 releases from Heavy Water Reactors.

A. BIRKHOFER, Germany

Let me comment on this question as co-author of the paper. For me optimisation and ALARA means about the same. Based on many years of experience in the licensing of nuclear power plants in Germany, I never found a strong argument from the industry against a further requirement to reduce the releases. I don't therefore agree with Mr. Berry when he says that the costs are extraordinarily high for the type of effluent systems we have today.

In our licencing procedure the applicant has to demonstrate that he meets the conditions of the radiation protection regulations (300 μSv annual whole body dose to the population). In doing so a set of conditions have been put which are very conservative, so in meeting this it turns out that during normal operation the individual doses are less than 10 μSv per year. That is just the consequence of a stringent set of rules that leads to those very low doses and in this respect it was in accordance with the ALARA principle and I never had been told by the industry that this was an approach which really was very difficult to meet.

P.Y. TANGUY, France

J'ai noté que M. Clarke considère que l'optimisation n'est pas facile quand elle porte sur des événements probabilistes. J'y viendrai dans ma présentation de demain.

Mais je ne suis pas sûr que l'optimisation soit effective pour les cas plus faciles dans lesquels l'exposition est certaine. Je reviens sur le point évoqué par M. Berry : les effluents radioactifs des centrales. Le rapport donne un risque individuel maximum de 1 μSv/an, et précise : "on peut considérer qu'il y a optimisation" (p. 4). Plus loin, il est écrit que 20 μSv/an représente un risque si faible qu'il ne doit pas être pris en compte au niveau individuel. Ceci ne me paraît pas cohérent. Si l'optimisation n'a pas été appliquée, il faut le reconnaître.

D. BENINSON, ICRP

This discussion gives me an opportunity to clarify that optimisation is not minimisation. The reduction of exposures stops when the effort needed for further reduction is larger than the advantage obtained. Optimisation has been applied for the design of control system of releases in Boiling Water Reactors, Heavy Water Reactors and in other installations. The end result is not zero, but to go lower than that would require an effort which is larger than what you gain. In these cases you can apply the cost-benefit analysis, because you are dealing with a single variable.

The difficulty of using optimisation comes in the case of accidents due to the multiple attributes involved (acute effects, late effects, land contamination, etc.), and which are not easily combined into a single quantity. We will come back to this question later.

R.H. CLARKE, United Kingdom

Several of the speakers have implicity raised the question about whether collective dose is a meaningful concept when the individual doses are very small. Mr. Tanguy's contribution was essentially saying that we should ignore small doses to a large number of people. I hope we will have the opportunity to discuss this point this week.

A.J. GONZALEZ, IAEA

One of the main problems between the radiation protection community and the nuclear safety community is that of intercommunication; they do not use a common thesaurus. A same term sometimes means different concepts for each community. I did not find this problem solved in this leading paper by two leading members of both communities. e.g. the term risk is used to mean both (i) the probability of harm (i.e. "radiation-protection" meaning) and (ii) the mathematical expection of harm (i.e. "nuclear safety" meaning). I would like to provoque both authors in providing us - in a leading paper - with a common thesaurus that we very much need to improve our common understanding.

THE PROBLEMS OF ESTABLISHING A BIOLOGICAL BASIS
FOR THE ASSESSMENT OF RADIATION RISKS

G. Silini
Italy

ABSTRACT

This paper reviews recent data about the deterministic, late somatic and hereditary effects of radiation. It also discusses the criteria needed to derive from the original information generalised risk assessments on which to found a system of radiation protection, the adjustments required to this end and the current risk evaluations. In the author's opinion, the estimates of deterministic effects are scientifically well founded; the projections of cancer risk are acceptable, since they originate from epidemiological data through reasonable assumptions and quantitative models; the assessments of hereditary detriment are weak, because they are derived from data on experimental animals with inadequate knowledge of basic human genetics. In all cases, however, these estimates are the best possible at present.

PROBLEMES SOULEVES PAR L'ETABLISSEMENT D'UNE BASE BIOLOGIQUE
PERMETTANT D'EVALUER LES RISQUES DUS AUX RAYONNEMENTS

RESUME

La présente communication fait le point des données récentes relatives aux effets déterministes, somatiques tardifs et héréditaires des rayonnements. Elle contient aussi un examen des critères nécessaires pour tirer, à partir des informations originelles, des évaluations généralisées des risques sur lesquelles un système de radioprotection puisse s'appuyer, des adaptations requises à cet effet et des évaluations actuelles des risques. De l'avis de l'auteur, les estimations des effets déterministes sont scientifiquement bien fondées ; les projections de risque de cancer sont acceptables, car elles ont été obtenues à partir de données épidémiologiques grâce à l'utilisation d'hypothèses et de modèles quantitatifs raisonnables. Les évaluations du détriment héréditaire sont, quant à elles, peu solides, puisqu'elles sont tirées de données découlant de l'expérimentation animale s'accompagnant d'une connaissance insuffisante de la génétique humaine fondamentale. Dans tous les cas, cependant, ces estimations sont les meilleures que l'on puisse obtenir à l'heure actuelle.

1. INTRODUCTION

I have been asked to present a frank discussion of the problems encountered when assessing radiation risks in a quantitative manner. I will address
this subject in a very simple language, discussing first the deterministic
(previously called non-stochastic) effects, then the stochastic somatic effects
and finally the hereditary effects. In adopting this arrangement I have implicitly decided to go from the easiest to the most difficult assessment.

There is no need to elaborate on the need that risk estimates for human
radiation protection should, if possible, be derived from experience on man.
Obviously, then, the scientific reliability of the assessments will depend on
the amount and quality of human information available, which decrease from the
deterministic, to the stochastic to the hereditary effects. For irradiation of
single tissues there is abundant data from the large number of people treated
every day in radiotherapy; accidental or medical exposures have also produced
useful information on the irradiation of the whole body; and one may fairly
confidently fill the gaps with research on experimental animals. Thus, for
deterministic effects we stand on rather firm ground. For cancerogenesis in
man, one has a number of epidemiological series, but shortcomings of different
kind render the assessments of neoplastic effects considerably more difficult;
furthermore, the mechanisms and rates of tumor induction cannot easily be ex-
trapolated between species. Thus, predictions of cancer risk are less reliable. As for hereditary effects, there is virtually no human experience and
one is totally dependent on animal data, corrected to some extent by information on human genetics. Consequently, assessments of the hereditary detriment
are at present unsatisfactory.

Considering now the practical issues, deterministic effects are of
interest only for irradiation at high doses and dose rates, which means (for
the purpose of this seminar) accidental exposures. Although accidental exposures will indeed be part of our discussions, they are by no means the most
common exposure situation for workers and population at large, for whom irradiation at low doses and dose rates is the rule. Since we assume, on the
contrary, that the probability of inducing stochastic effects, both somatic and
hereditary, is increased even for the smallest doses, these effects are a cause
for concern under all exposure conditions.

Thus, it seems reasonable to discuss the deterministic effects rather
shortly, to give more attention to the stochastic ones.

2. IMMEDIATE SOMATIC (DETERMINISTIC) EFFECTS

2.1 Post-natal irradiation

The immediate somatic effects are expressed in man within a few days or
weeks from exposure and are caused by damage to one or more of the self-
renewing tissues of the body, with consequent loss of specific function. All
immediate effects depend on the inactivation of a great number of functional
cells of a given differentiating line, for example, the cells of the skin, the
white blood cells, the intestinal cells, and so on. These effects may be

localised in a given tissue for partial-body irradiation or generalised for whole-body exposure, often under the form of a specific syndrome, i.e., a cluster of symptoms characteristic of a given dose range [1].

Immediate effects on tissues are rather easily recognised to an expert eye: they are deterministic in nature, meaning that they are expected to occur with a given probability and degree of severity in individuals irradiated to a given dose. If one plots the probability of induction of any deterministic effect against dose, one finds a sigmoid curve: probability is zero up to a threshold dose of a few hundred mSv for the most sensitive tissues; then it rapidly increases to unity, the slope reflecting the spread in sensitivity within the irradiated population. Each effect has a characteristic threshold, for a given irradiation regime. Above the appropriate threshold, the severity of the effects also increases with dose and usually with dose rate. The dose rate dependence occurs because spreading the dose in time allows part of the radiation injury to be repaired, through various mechanisms operating at the cellular or tissue level [1].

In the context of this seminar it is also appropriate to mention short-term radiation lethality, an item that one may neglect when discussing small doses but may become important for accidental over-exposures. This effect is expressed as LD_{50} (i.e., the dose reducing survival to 50%) at 1 or 2 months post-irradiation. In its 1988 report, the UNSCEAR [2] reviewed radiation lethality in man and identified a fairly large spread of LD_{50}'s for groups of people who were in different states of health at irradiation and received different degrees of medical attention thereafter. As a result of this re-analysis, UNSCEAR no longer recommends a single value of LD_{50} in man, but rather a range from 2.5 to 5 Gy (bone-marrow dose). Consequently, for planning of countermeasures it becomes important to know which value of the LD_{50} might be appropriate to choose for the accidental situation envisaged.

For radiation protection purposes, the existence of a dose threshold is the most important feature of deterministic effects, because it allows to avoid them completely by keeping the absorbed dose below the appropriate threshold. Thus, under normal conditions of exposure, the immediate somatic effects of radiation are easy to deal with and are not supposed to arise for irradiation below the limits. It is also important to remember that threshold values usually assumed in radiation protection apply to low doses and dose rates, whereas thresholds for single brief exposures are usually lower. This must be accounted for when dealing with the consequences of accidents.

Deterministic effects resulting from densely-ionizing radiation are similar in nature, but their frequency and severity per unit dose is higher - and their threshold lower - than for sparsely-ionizing radiation. If one compares the effectiveness of the two types of radiation, the maximum values of the effectiveness for deterministic effects on many tissues are 2-5 times lower than for stochastic effects in the same tissues [3]. Therefore, the radiation weighting factors (w_r) defined for stochastic effects overestimate the contribution of densely-ionizing radiation for immediate somatic effects.

2.2   Pre-natal irradiation

A special class of deterministic effects is that induced on the conceived but unborn child. The outcome of irradiation in utero of the embryo and

fetus depends very strongly on the developmental stage. If radiation is absorbed by an early embryo when differentiation is not very advanced, the loss
of even a few cells may result in failure of the embryo to implant into the
uterine wall: this will cause embryonic death, an effect which may often go
clinically undetected. During the phase of major organogenesis, irradiation
may induce malformations in the anatomical structures that are developing at
the time of exposure. There is little known in man, but in experimental
animals malformations have dose thresholds of 0.05-0.1 Gy, increasing with the
stage of development. As time proceeds and the fetus continues to develop,
growth retardation may also set in, but only at fairly high doses [4, 5].
Malformations and growth retardation are not very critical in the context of
radiation protection, except for accidental over-exposures.

An interesting outcome of human irradiation in utero is an alteration of
the brain cortex architecture, occurring between 8 and 25 weeks from conception. At 8-15 weeks the precursors of neurons are generated inside the brain
hemispheres and then migrate to the cortex; at 16-25 weeks the mature neurons
establish connections between themselves and with glial cells, in an elaborate
network which is typical of the central nervous system. Disruption by radiation of this precisely coordinated process results in variable degrees of mental retardation in the exposed child, manifested as a dose-related decrease of
cognitive functions (IQ scores, delay in the appearance of the major landmarks
of physical and intellectual development, impaired school performance). Experience on atomic-bomb survivors shows a radiation-induced shift in the normal
IQ distribution of about 30 IQ points per Sv for exposure at 8-15 weeks [6].

In extreme cases, the damage of the brain cortex may produce very
severe mental retardation, with a probability of occurrence of 45 percent per
Sv at 8-15 weeks and 10 percent per Sv at 16-25 weeks. Human experience for
this effect is limited to about 30 cases of retarded children out of about
1 000 cases exposed in utero. For such an effect, a threshold of a few hundred
mSv may not be excluded, and this would make it unlikely for planned exposures.
Note that these estimates of risk are about a factor of 10 higher than for
stochastic effects and refer to single acute irradiation, which again may only
be of interest for accidents [6].

3.   LATE SOMATIC (STOCHASTIC) EFFECTS

Late somatic effects are expressed as extra cases of leukaemia and
cancer in an irradiated population: these are absolutely indistinguishable
from similar malignancies occurring in the absence of irradiation above natural
background. Late effects are stochastic, because they occur at random in the
exposed population and it is impossible (at least up to now) to identify in any
single case any correlation with radiation exposure. The causal relationship
between irradiation and induction may only be shown over large populations, as
a statistically significant increase of cancer and leukaemia over the natural
incidence.

The epidemiological series available to derive risk estimates for radiation protection are fairly numerous and include people irradiated for a variety

of reasons such as military attacks; medical, occupational and accidental exposures; exposures in regions of high natural background. However, to compound all information into coherent estimates of risk is very hard for one or more of the following reasons:

a) The uniqueness of some series;

b) The variable methodologies adopted (prospective versus retrospective studies, case-control versus cohort studies);

c) The differences in size and duration of the series;

d) The uncertainties of dose and dose-distribution assessments;

e) The frequent lack of homogeneous control groups [2].

Since it is impossible to correct the intrinsic characteristics of the various studies, the only practicable course of action is to adjust the results onto a common pattern, corresponding to the exposure conditions for which the risk of cancer is relevant, that is the low doses (10-100 mSv) delivered at low dose rates (tens of mSv per year). However, conceptual problems make such an objective even harder. Within the time available I can only discuss the major issues, referring mainly to the 1988 UNSCEAR report [2].

3.1 The risk over organs

The great majority of studies concerns single organs and tissues: the risk coefficients obtained from these series are very scattered, presumably for the reasons already cited. There are two ways to overcome this variability: either to combine the estimates disregarding the differences between series, a procedure scientifically objectionable; or to focus attention upon those series where most organs and tissues were irradiated together, in the hope of obtaining more homogeneous results. The UNSCEAR followed this latter procedure: it compared the results obtained on the survivors of the atomic bombs and on the ankylosing spondylitis and cervical cancer patients, after making sure that the data from these three series were broadly in agreement with those of all other studies on individual organs and tissues. The results of the comparison are given in Table 1 and show a good agreement in identifying the most susceptible organs and tissues, with some scatter of the relative risk between the three series, and particularly in the last one. Radiobiological considerations may to a large extent explain the scattering of the data, but the variability may largely be overcome by pooling the results into two classes of malignancy: leukaemias and all other tumours. This is scientifically not very justifiable, but useful in practice.

3.2 The risk over time

In many of the ongoing studies, several decades after irradiation people still show an excess of some tumour types. There is no way to know when the appearance of extra malignancies will come to an end and what will eventually be the final score of cancers among irradiated people; on the other hand, ignoring this future component of risk would not offer complete protection. Thus, one must project through mathematical models the cancer experience of people who died already to the likely future experience of those who are still alive.

Table 1

SUMMARY OF THE ESTIMATED RISK OF CANCER PER 1 GY OF ORGAN ABSORBED DOSE
OBTAINED FROM THE ATOMIC BOMB, ANKYLOSING SPONDYLITIS
AND CERVICAL CANCER SERIES (SIMPLIFIED FROM [2])

EXCESS RELATIVE RISK

Organ or tissue	Atomic bomb survivors	Spondylitis series	Cervical cancer series
Leukaemia	5.21	3.5	0.88
All cancers except leukaemia	0.41	0.14 *	**
Bladder	1.27	0.19	0.07
Breast	1.19		0.03
Kidney	0.58	0.12	0.71
Large intestine	0.85		0.00
Larynx	0.51	0.15	
Lung	0.63	0.13	
Multiple myeloma	2.29		
Oesophagus	0.58	0.29	
Ovary	1.33	0.00	0.01
Rectum	0.00	0.03	0.02
Stomach	0.27	0.004	0.69

ABSOLUTE RISK
(excess deaths per 10^4 person–year–Gy)

Leukaemia	2.94	2.02	0.61
All cancers except leukaemia	10.13	4.67	**

* All cancers, except leukaemia and colon cancer.
** The estimate cannot be made for this series. An estimate of the whole-body dose does not exist, and probably cannot be obtained given the nature of the exposure.

Two risk projection models may be used to this end: the absolute or additive model, postulatinig that the number of extra tumours expressed by an exposed population will only be a function of dose; and the relative or multiplicative model, which assumes that the tumours expressed will be both a function of the dose and of the natural frequency. The latter model seems at present to describe the data from the atomic-bomb survivors better than the additive model. Reasonable projections may be made with both models on the basis of present risk coefficients on populations of a given sex and age structure, for which the time evolution and the natural cancer incidence are known. The additive model calculates less tumours than the multiplicative one, but with an earlier onset in time. This is shown in Table 2, which gives the projected number of malignancies and lifetime lost by the two models in an adult population of 500 males and 500 females exposed to 1 Gy of high-dose-rate low-LET radiation.

Table 2

PROJECTION OF LIFETIME MORTALITY AND LOSS OF LIFE EXPECTANCY
FOR AN ADULT POPULATION OF BOTH SEXES (500 MALES AND 500 FEMALES)
EXPOSED TO 1 GY OF ORGAN ABSORBED DOSE OF SPARSELY-IONIZING RADIATION
AT HIGH DOSE RATE. DATA REFERRING TO THE JAPANESE POPULATION
(SIMPLIFIED FROM [2])

MALIGNANCY	EXCESS FATAL CANCER		YEARS OF LIFE LIST	
	multiplicative	additive	multiplicative	additive
Leukaemia	8.6	10	130	230
All cancers except leukaemia	47	36	490	610
TOTAL	56	46	620	840

3.3 The risk over age

The most recent findings in Japan show that the risk of developing cancer is higher in survivors who were exposed under 20 years of age. The relevant data are still rather few and have large errors, because those exposed as children have not yet entered the age of high cancer incidence: more precise estimates will be possible in the future. For risk projection purposes, this phenomenon has a radiobiological aspect, consisting in an intrinsically higher susceptibility to cancer of the younger age cohorts; and a demographic aspect, due to the fact that young people have a longer expectation of life and, therefore, more time to express the consequences of exposure. There are various ways to calculate the relative contribution of the young ages in the whole population. As an example, Table 3 shows that the projected number of cancers and loss of lifetime is lower in an adult population than in one where all ages are represented, thus indirectly pointing to the higher risk of the young people.

Table 3

PROJECTION OF LIFETIME MORTALITY AND LOSS OF LIFE EXPECTANCY FOR A POPULATION OF BOTH SEXES (500 MALES AND 500 FEMALES) AND ALL AGES EXPOSED TO 1 GY OF ORGAN ABSORBED DOSE OF SPARSELY-IONIZING RADIATION AT HIGH DOSE RATE. DATA REFERRING TO THE JAPANESE POPULATION (SIMPLIFIED FROM [2])

Population type	Projection model	Excess fatal cancer	Years of life lost
Total population	Additive	40– 50	990–1200
	Multiplicative	70–110	950–1400
Population over 25 years	Additive	50	840
	Multiplicative	60	640

3.4   The effect of dose rate

Most epidemiological series refer to irradiation at high doses and dose rates, while in practice people are mostly exposed at very low doses and dose rates. It must be realised that what determines the real added risk due to an increased exposure to radiation is the actual form of the dose-response relationships for cancer induction. Radiation biology conclusively shows that for sparsely-ionizing radiations this form is rarely linear for complex effects in mammals. In recent times a linear-quadratic model has frequently been fitted to experimental and human data, such as,

$$E = a D + b D^2$$

in which the effect, E, is dependent on the sum of two component terms, one linear and one quadratic with the Dose D, the constants a and b being the numerical expression of these components.

In this type of relationship the constant a prevails at low doses, while b tends to prevail with increasing doses and dose rates. To put some value to these coefficients, one might say that b becomes absolutely predominant above about 1 Gy or 1 Gy/min. although the actual values of a and b change in absolute terms and relative to one another in different experimental and epidemiological series [5].

It follows from this discussion that when one tries to infer the value of the linear term from data at high doses where the quadratic component is operating, one is likely to overestimate the slope of the linear part. It is therefore common practice to reduce the coefficients obtained from observations at high doses and dose rates. From a recent discussion of this subject, UNSCEAR [2] concluded that – in various experimental models and for single tumour types – reduction factors of between 2 and 10 could be derived. In Hiroshima, a factor of 2 could be justified for leukaemia, but for all solid tumours considered together linearity seems at present to prevail. The whole subject needs re-examination. For the moment, the ICRP divides by 2 all cancer risk estimates obtained at or above 0.2 Gy with dose rates of, or higher than, 0.05 Gy per hour [7].

3.5   Past and present risk estimates

Owing to all these and other difficulties, it is hard to derive gener-alised estimates of the probability of tumour incidence per unit dose. This is done by considering all information from many epidemiological series and by averaging over them as best as one can. The derived values are called nominal risk coefficients to mean that they are different from those obtained from real populations irradiated with defined exposure patterns. For leukaemia and can-cer together, the estimates of the nominal risk coefficients obtained since 1977 by various evaluators are given in Table 4. They include all the differ-ent data sets, projection models, extrapolation assumptions, population types, dose and dose-rate reduction factors applied by the different evaluators at various times.

Table 4

RISK OF INDUCTION PER UNIT DOSE OF ALL MALIGNANCIES IN THE GENERAL POPULATION, AS CALCULATED AT VARIOUS TIMES BY VARIOUS EVALUATORS

EVALUATOR AND DATE	RISK (per Sv)	
	High dose rate	Low dose rate
UNSCEAR, 1977 [4]	2.5	1.0
ICRP, 1977 [8]		1.25
BEIR III, 1980 [9]	1.5*	0.8*
NUREG, 1985 [10]	2.0*	
UNSCEAR, 1988 [2]	4.0–11.8	
BEIR V, 1990 [11]	7.9*	5.6*

* Central values, see Table 4.4 in [11].

It is shown that over the last few years there was a distinct increase in the estimates of radiation-induced malignancies per unit dose, to which the following main factors have contributed in variable amounts:

a) The continued appearance of malignancies in further follow-up of the ongoing series;

b) A downward revision of the dosimetry in Japanese survivors, leading to a corresponding increase of the estimates;

c) The recent application of multiplicative projection models, instead of the additive models previously in use;

41

d) The realisation of the higher susceptibility of the younger age
cohorts.

Based on present knowledge and taking all above points into account, the
ICRP proposes now [7] to take the nominal lifetime fatality probability coefficient for all malignant conditions at 4 percent per Sv (low-dose-rate,
sparsely-ionizing radiation) for people of working age. The same coefficient
for the general population is taken to be 5 percent per Sv. The mean loss of
life per attributable malignancy is assumed to be 20 years for the additive and
13 years for the multiplicative model.

3.6   Applicability of the risk estimates

There remains one last important point to consider: to what extent are
the risk estimates in Table 4 applicable to the whole human species, since they
have been derived mostly from populations where radiobiological and demographic
data happened to be more abundant. This problem was addressed in the 1988
UNSCEAR report [2], which compared the results of applying the risk coefficients for leukaemia and all other malignancies obtained from Japan to the populations of Japan, the United Kingdom and Puerto Rico, having different baseline mortality patterns. Projections by the additive model were similar, while
the multiplicative projections differed by 20% as a maximum. The exercise
showed that such projections were rather insensitive to differences in overall
and cancer-specific mortality pattern, but did not exlude much larger differences in organ-specific risk.

A more refined exercise by the ICRP Committee 1 [7] confirmed and
enlarged these conclusions. It showed that, depending upon the transfer and
projection models used, differences between organs are within a factor of 2-3,
but the overall risk estimates differ by less than 30% of each other and by
about 10% from the average. The variability of the risk coefficient itself
between different ethnic groups remains for the moment an unresolved point.

4.   HEREDITARY (STOCHASTIC) EFFECTS

Hereditary effects are those induced in the progeny of irradiated individuals. They may appear within the first generation from irradiation (dominant); or within subsequent generations, when the same two mutations, one from
the father and one from the mother, match into the progeny (recessive damage).
In the genetic material radiation damage may take the form of alterations of
small parts of the genome or of entire chromosomes. There is little corre-
spondence between this damage and the severity of the resulting clinical conditions: thus, individuals carrying the most devastating cellular alterations
may never be born, while the apparently less serious gene mutations may result
in severely handicapped progeny.

In viable children, some hereditary diseases are rather mild and allow
an essentially normal life (imagine, for example, a condition like colour
blindness), while others are very serious (mongolism, for example). Radiation
protection usually makes reference to the most severe conditions which are
either incompatible with life or very disabling for the individual affected and

heavy on his family and the whole society. There is no way to distinguish between the radiation-induced conditions and those which appear naturally in a very large proportion of new-born individuals. It is assumed that hereditary diseases will appear randomly in the progeny of irradiated subjects. Up to now, however, there is no clear demonstration that radiation may indeed cause hereditary effects in man, although there is ample evidence for them in many other species and no reason to doubt that, at appropriate doses, they may also be induced in man.

4.1   Methods for assessing hereditary risk

For single-gene mutations and chromosomal aberrations, radiation protection assumes a linear non-threshold dose-induction relationship. On this assumption, two basic methods are currently in use for hereditary risk assessment [2]:   the direct method, roughly equivalent to the additive risk projection model described for cancer;   and the doubling-dose method, similar to the multiplicative model. With the former method one derives the risk of dominant mutations from similar data on the mouse, and the risk of chromosomal aberrations from data on monkeys. Primary data are then corrected to account for radiosensitivity, germ cell stages, dose and dose-rate relationships and relative viability of the various types of aberrations in animals and man.

With the doubling-dose method one starts from the rate of naturally-occurring disorders in man, and derives from low-dose-rate data in the mouse the dose needed to double the natural rate at equilibrium. From these two numbers, under certain assumptions, one can calculate the increased probability of induction of genetic disorders per unit dose at equilibrium and the probability of disorders in the first generation.

The most important assumptions involved in the use of the doubling-dose method are that:

a) At any given time there is an equilibrium in the population between newly-arising mutations and mutations eliminated by natural selection;

b) At a higher irradiation level the population will eventually reach a new and higher equilibrium;

c) There is proportionality between the number of new mutations expressed over all generations by a single exposure and that expressed at equilibrium under continuous exposure at the same dose per generation;

d) The mutation rate is similar in males and females;

e) There is proportionality between mutation and disease;

f) The spectra of spontaneous and induced mutations are similar.

All these assumptions are plausible within the theory of population genetics, but one ignores to what degree they might apply to man.

As if all this were not enough, there is also another major problem. Most genetically-determined conditions occurring naturally in man are multifactorial, i.e. dependent on a number of altered genes. Multifactorial conditions are maintained in part by the occurrence of new mutations in the population, and in part by environmental factors. It is impossible at present to assess the relative importance of these two components: consequently, radiation-induced multifactorial disorders cannot be confidently quantified [2].

4.2 Past and present estimates of hereditary risk

Estimates of the hereditary risk of irradiation by some independent evaluators have changed since 1972, as shown in Table 5. The apparent slight tendency to a decrease over the years is mostly due to the exclusion of multifactorial disorders that cannot be assessed at present.

Table 5

NUMBER OF SEVER HERDITARY CONDITIONS EXPRESSED OVER 1.000.000 LIVE BORN
ESTIMATED BY THE DOUBLING DOSE METHOD, AS CALCULATED BY VARIOUS EVALUATORS.
THE ASSUMPTION IS THAT BOTH PARENTS OF THESE CHILDREN
RECEIVED 1 GY PER GENERATION OF SPARSELY-IONIZING RADIATION
DELIVERED AT LOW DOSE RATE (DATA FROM [2])

EVALUATOR AND DATE	DOUBLING DOSE(Gy)	NATURAL PREVALENCE	RADIATION-INDUCED INCREASE	
			1st gen.	all gen.
UNSCEAR, 1972 [12]	0.2-2.0	60.000	10-200	60-1500
UNSCEAR, 1977 [4]	1	105.100	63	185
BEIR III, 1980 [9]	0.5-2.5	107.100	15-75	60-1100
ICRP, 1980 [13]	1		89	320
UNSCEAR, 1982 [1]	1	105.900	22	150
NUREG, 1985 [10]	1	50.900	30	185
UNSCEAR, 1986 [5]*	1	163	18	104
UNSCEAR, 988 [2]**	1	130	18	120
BEIR V, 1990 [11]***	1	360-460	15-40	115-215

*   Excluding multifactorial
** Excluding multifactorial and numerical chromosomal
*** Excluding multifactorial, including congenital abnormalities

Currently, the ICRP proposes [7] to take the nominal risk coefficient for quantifiable serious hereditary effects over all generations, relative to the dose absorbed into the gonads, at 0.5 per cent per Sv. Assuming that multifactorial disorders may contribute an equal amount of detriment, the whole risk rises to 1 per cent per Sv.

Obviously, the damage to the gonads of people too old to produce children will never be expressed and, therefore, only the doses absorbed by individuals of fertile ages will give rise to hereditary effects. It is estimated that the nominal risk coefficient for genetic effects for irradiation of a population of workers will only be about 60% of that applying to the population as a whole, that is 0.6 per cent per Sv [7].

5.   CONCLUSIONS

In conclusion, whilst problems for a quantitative assessment of deterministic effects are, in general, not too difficult to solve, one should always remember that the radiobiological mechanisms for small doses and low dose rates are different from those operating at high doses and dose rates. Any realistic system of radiation protection should allow for these differences and recommend actions and interventions appropriate for the exposure conditions to be faced in practice.

Assessment of risk coefficients for late somatic (stochastic) effects is possible, but requires corrections of the primary risk factors obtained from human epidemiological series to account for various dependencies of the effects on dose, dose rate, organs, age, time and other variables of radiobiological interest. Recently, understanding of all these risk-related variables has considerably increased and, as a consequence, one feels that current risk estimates are more realistic than in the past and less likely to increase further, barring drastic changes in the trend of the primary data.

Estimates of the hereditary risk are largely based on experimental animal data and dependent on a number of important assumptions still to be validated in man. The occurrence of hereditary effects has not even been demonstrated in the human species, but this is not in itself a reason for reassurance nor concern. Assessments of the hereditary risk have shown a slight apparent reduction over the years, due more to a decrease in our confidence to quantify part of the damage than to a real reduction of the overall estimates. Thus, future progress will critically depend on advancement of basic human genetics.

6.   REFERENCES

[1]   UNSCEAR. Ionizing Radiation: Sources and biological effects. United Nations, New York, 1982.

[2]   UNSCEAR. Sources, Effects and Risks of Ionizing Radiation. United Nations, New York, 1988.

[3]   ICRP. RBE for Deterministic Effects. ICRP Publication 58. Annals of the ICRP, 20, No.4, 1989.

[4]   UNSCEAR. Sources and Effects of Ionizing Radiation. United Nations, New York, 1977.

[5]   UNSCEAR. Genetic and Somatic Effects of Ionizing Radiation. United
        Nations, New York, 1986.

[6]   ICRP. Developmental Effects of Irradiation on the Brain of the Embryo
        and Fetus. ICRP Publication 49. Annals of the ICRP, 16, No.4, 1985.

[7]   ICRP. Recommendations of the Commission. Draft February 1990.
        ICRP/90/G-01.

[8]   ICRP. Recommendations of the International Commission on Radiological
        Protection. ICRP Publication 26. Annals of the ICRP, 1, No.3, 1977.

[9]   NRC, Committee on the Biological Effects of Ionizing Radiations. The
        Effects on Populations of Exposure to Low Levels of Ionizing Radiations
        (BEIR III). National Academy Press, Washington, 1980.

[10]   Evans, H.G., Moeller, D.W. and Cooper, D.W. Health Effects Model for
        Nuclear Power Plant Accident Analysis. NUREG/CR 4214, 1985.

[11]   NRC, Committee on the Biological Effects of Ionizing Radiations. Health
        Effects of Exposure to Low Levels of Ionizing Radiation (BEIR V).
        National Academy Press, Washington, 1990.

[12]   UNSCEAR. Ionizing Radiation: Levels and Effects. 2 vol. United
        Nations, New York, 1977.

[13]   Oftedal P and A.G. Searle. An overall genetic risk assessment for
        radiological protection purposes. J. Med. Genet. 17, 15-20, 1980.

THE PROBLEMS OF ACHIEVING NUCLEAR SAFETY OBJECTIVES

H.J.C. Kouts
Defense Nuclear Facilities Safety Board
United States

ABSTRACT

The problems inherent is setting safety goals and in assessing their achievement are discussed. A difficulty underlying all such actions is the need for probabilistic safety assessments, which are not as widespread as they need to be for use of safety goals. The difficulties associated with use of PSA are discussed. Other difficulties in achieving goals stem from the age of some plants, so that they do not have all modern recognized hardware features, and the resistance of some managements to change. The problems of assessing the meeting of safety goals are discussed.

LA REALISATION DES OBJECTIFS EN MATIERE DE SURETE NUCLEAIRE ET LES PROBLEMES QU'ELLE SOULEVE

RESUME

Les problèmes inhérents à la fixation des objectifs en matière de sûreté et à l'évaluation de leur réalisation sont examinés dans la présente communication. Ces démarches se heurtent toutes à une difficulté, à savoir la nécessité de recourir à des évaluations probabilistes de la sûreté (EPS), lesquelles ne sont pas aussi répendues qu'elles le devraient pour la mise en oeuvre des objectifs de sûreté. Les difficultés liées à l'utilisation des EPS font l'objet d'un examen. La réalisation des objectifs se heurte à d'autres difficultés imputables à l'âge de certaines installations, qui fait qu'elles ne sont pas dotées de tous les équipements modernes éprouvés de sûreté, et de la résistance au changement de certains dirigeants. L'auteur aborde les problèmes que pose le respect des objectifs en matière de sûreté.

GENERAL REMARKS

Although safety goals or objectives have not been adopted universally for nuclear power in countries that possess these plants, they are in place in many countries. They have been formulated because in political circles it has been suggested that regulators of nuclear plant safety must have been guided in their actions by views as to how safe they believe the industry should be. This observation is often stated as: how safe is safe enough?

Adoption of goals of this kind implies some prior thinking as to whether and how they are to be achieved, and then an ability to be able to determine whether the objective has been met; otherwise, the exercise would be empty of meaning. Achievement of the goals requires a combination of activities addressing both hardware and practices. Actions of this kind have been under way since the beginning of the nuclear power industry, with the hardware having received most of the attention until recently. Development of the ability to determine whether safety goals have been achieved is, however, an even more difficult task than settling on the goals themselves or the means to address them.

THE REQUIREMENT FOR PSA

Before we consider these questions of the implementation of safety goals, it would be useful to discuss the basis for the process itself. The concept requires an appreciation of the fact that safety is not absolute, but that it is a relative thing that can even be quantified. There is such a thing as less safety and another thing that can be called greater safety, but there is no such thing as safety. To adopt the concept of safety goals is to accept the need for probabilistic safety assessment, because only PSA can provide the quantification of safety that is necessary. PSA was invented for precisely this purpose. This is important because many people cannot understand or accept this starting point. Safety goals as we are discussing them have no importance or relevance to people who seek absolute safety. They will assume when we discuss safety goals that we are only temporizing and hiding the fact that nuclear plants are unsafe. We must realize that we are instead speaking to another audience which is more sophisticated and can accommodate to quantitative thinking.

THE NATURE OF SAFETY GOALS

The safety goals that have been developed in different countries are not uniform in appearance. In the United States, the Nuclear Regulatory Commission has adopted qualitative goals that address the possible consequences of accidents to nuclear power plants. They are meant to ensure that over the possible history of use of nuclear power in the United States, accidents to nuclear plants will not contribute appreciably to causes of loss of life either on a collective or an individual basis. They are meant to ensure that nuclear plants are among the most benign means of producing electricity. In most other countries, goals

are directed to avoidance of severe accidents.   The goals INSAG
has stated in its document, <u>Basic Safety Principles for Nuclear
Power Plants</u>, fall between.   They seek to limit the probability
of a severe accident and also the specific consequences that
could lead to a need for measures offsite to protect the
neighboring population and the environment.

    It is important to note that none of the goals voiced in
different places is couched in terms that imply that accidents
take place regularly.   In this respect they differ from safety
goals that might be set for air travel, mining, flour mills, or
almost any other industry.   The two accidents that have destroyed
nuclear power plants are so rare as to fail in establishing a
rate at which others may be expected.   This is the reason for the
need for PSA.

DIFFICULTIES INHERENT IN USE OF PSA

    The methods and results of probabilistic safety assessment are
important both to meeting safety goals and assessing how well
goals may have been met.

    It is now recognized that the most important application of
PSA is to determining those sequences that contribute most to
risk to the plant or the population or environment about it.
These sequences define the vulnerabilities of the plant.   They
indicate the measures that can have the greatest benefit in
reducing risk.   In principle, this method can shed light on both
hardware and procedural issues.   Of course, the contribution of
each sequence to risk has large associated error margins, but it
is commonly assumed that the relative values of contributions to
risk are more reliable,

    The application to determining how well goals have been met
can be done either to individual plants or to a number of plants
comprising a nuclear industry.   The application to individual
plants is usually discouraged, because the large error margins
attached to results of PSA's sometimes leads to indefinite
conclusions as to the meaning of the comparison of the safety
goal to the calculated result.   Yet the temptation to use the
results of the PSA in this kind of bottom line comparison can be
overwhelming.   The comparison certainly has the benefit just
described, of illuminating the ways by which risk may be reduced.
Beyond this, it even helps in determining which nuclear plants
stand more in need of improvement than others, because the use of
a common data base causes relative values of risk between two
plants to be more reliable than the risks of the individual
plants.

    The most widespread obstacle to this important application of
PSA to meeting safety goals and to determining how well they are
met is the uneven state of probabilistic safety assessment
throughout the world.

At least nine distinctly different types of nuclear power reactors are in use throughout the world. About 350 of the approximately 450 plants are water reactors of the pressurized or boiling types. There is an extensive set of probabilistic safety assessments relative to plants in this category, and the quantity of them is constantly increasing. Few thorough PSA's have been done on any of the other reactor types. Where this deficiency exists, there is no ability to test the safety of these plants against safety goals.

Furthermore, there are problem areas in the PSA's of water reactors. These reactors have a variety of designs, especially in the United States, where almost every nuclear utility has its own plant designs, and sometimes a single utility has plants of several designs. The problem has been the result of the existence of four suppliers of nuclear steam supply systems, and more than half a dozen separate engineering concerns serving as the designers and contractors for the balance of plant. The result has been approximately a hundred separate nuclear power plant designs in the United States. Superposed on this variability are the differences in choices of components for essentially the same function, components such as valves, pumps, switchgear, piping, electrical circuitry, etc.

Most of the separate PSA's for water reactors are for plants in the United States, where the differences make it difficult to pull together a common data base or to transfer data from the analysis of one plant to another. The result is a data base whose members describe relatively broad classes of components and systems rather than specific designs. Therefore the uncertainty margins for failure rates determined from the data base are wider than if they referred to specific designs.

Other factors also contribute to error margins of PSA's and complicate the application of these methods to meeting safety goals. Some phenomena important to the safety analysis are still poorly understood, especially the physical and chemical processes that would take place after severe damage to the core of a nuclear plant had begun. This source of uncertainty affects use of safety goals such as those in use in the United States. But the greatest source of uncertainty in a PSA is surely the imperfect understanding of human reliability and its effects. These effects could be felt either through errors of omission and commission that might contribute to severe damage to a nuclear plant or the consequences thereof, or through successful recovery from a sequence of events that could have such consequences. A number of types of methodology are in use for generating data on human reliability to be used in PSA's. Many of them were tested against each other in the Human Factors Reliability Benchmark Exercise reported by ISPRA last year. In this exercise a number of analytical groups from a number of facilities independently calculated the recovery probability in a well-defined situation, and arrived at results whose range covered several orders of magnitude. These difficulties are discussed in NUREG-1420, which is a review of NUREG-1150 by an international panel, just

published. It is not yet possible to foresee the development of
methods of estimating human performance having the same level of
reliability as that of well-understood engineering.

A related source of uncertainty in safety assessment is the
effect of safety culture and its variability from one plant's
operating organization to another's. These consequences of
differences in management style and personal attitudes are not
taken into account in any PSA, yet they must have a vital effect
on true risk.

Finally, there are difficulties in assessing the influence of
ceratin external events. In determining the ability of a nuclear
plant to remain undamaged at a probability of 10^{-5} per operating
year, it is necessary to take into account large earthquakes and
floods and storms whose occurrence would be equally rare.
Reliable quantitative records of these disasters have been
accumulated for only about a hundred years, less in some areas.
Extrapolation based on uncertain understanding of the phenomena
must be used instead.

Altogether, the dependability of probabilistic safety
assessment is an obstacle to confident application of safety
goals and assessing compliance with them. It is an imperfect
tool whose use is mandatory where safety goals are stated and
used.

MEETING SAFETY GOALS

The means by which safety goals are implemented are the
classic practices of nuclear plant safety. These have been
embodied in the INSAG document previously referred to. They rest
on a solid combination of deterministic and probabilistic
analysis that guides the choice among design options and the
practices used in operation of plants. The central features are
informed and intelligent application of the concept of defense in
depth, and an effective and vigilant safety culture. In issuing
this document, INSAG made a judgement that the practices
advocated in it would in their implementation lead to achieving
safety goals that were also stated.

Achieving safety goals about the world is severely handicapped
by the fact that many of the nuclear plants are ten or more years
old, and were designed using safety standards in use on the order
of twenty years ago. They do not usually embody all of the
safety features based on lessons learned in the more recent past.
It cannot be expected that these older plants are inherently able
to meet standards as high as those to which plants of the current
generation conform. This point was taken into account by INSAG
when it stated its safety goals: a probability of severe core
damage less than 10^{-4} per year for existing plants, and 10^{-5} for
future plants.

Where older plants cannot meet such goals, two options are available. One is to backfit corrective hardware, and the other is to adopt compensatory procedural remedies. It should be noted that the use of probabilistic safety assessment is inherent in deciding on either choice, through a specific assessment or the application of a generic requirement based at least in part on probabilistic arguments. The process is under way in the United States in a two step process. Older plants are already being subjected to a Systematic Evaluation Process, by which deficiencies are revealed from requirements applicable to current plants, and compensatory features or measures are developed. The other step is the requirement for an Independent Plant Examination for each nuclear plant to establish on an analytical basis its level of safety. This would commonly be a probabilistic safety assessment or an argument based on similarity or dissimilarity to other plants on which a PSA had been made.

All such measures to improve safety to meet stated goals must take advantage of some criteria on which the decision as to a need to improve is based. Where the decision rests on a generic requirement, the basis for change is clear. Where it is motivated by the results of a PSA, things are sometimes not so clear. Who is to say that a plant with a calculated core melt probability just on one side or the other of a stated safety goal must be modified? For this kind of decision a graded response has been suggested but not adopted. Probabilities exceeding the safety goal by some factor would require a remedy within a time interval depending on the factor. Factors below some de minimis value would establish that there is no need for change.

Where hardware changes are not possible or are prohibitively difficult or expensive, procedural fixes are in principle possible. They may involve reductions in operating parameters, limits on operating methods, new operating limits, new operating and emergency procedures, additional training of operating and maintenance staff, etc. This highlights the importance of procedural matters, such as the development of good operating procedures, the extension of good quality assurance practices (not after-the-fact paper trails) throughout the plant and its processes, rigorous procedures for training and qualifying operating and maintenance staff, the development of high morale, and in general the features of a good safety culture. These matters are discussed at length in a new INSAG publication, INSAG-4, which should be coming into print just about now.

Other difficulties in meeting goals at older plants are commonly rooted in attitudes of management. It has been a long and tedious process, particularly in the United States, to convince all utility management organizations that nuclear plants represent a higher level of technology than the fossil fueled plants in the electrical generating mix, calling for the cultural level that is needed for the higher level of safety. Where management has come to accept the need for distinctive measures at nuclear plants, they have often been slow to recognize and understand the meaning of the culture that is demanded. The problems are usually manifested in a resistance to change. It is defensively pointed out by the management and the staff that the old existing methods have been successful so far, because they have been used for a long time without any accidents having taken place, and it is said that the advocates of the new methods are inexperienced and impractical.

Steady pressure by the NRC and INPO have slowly diminished this island of resistance. Several groups of private consultants who spent their formative years in the nuclear navy where they learned conduct of operations at nuclear facilities have been working with one nuclear utility after another, restructuring what they found to be inadequate in conduct of operations. These measures altogether have been effective in improving the procedural aspects of operation at the older plants, to compensate for some of the departures from today's hardware requirements. Unfortunately, there is no good way to incorporate into the PSA's the effect of this method of trading procedural improvement for hardware deficiencies. Data for human factors are usually generic and do not reflect differences in management attitudes or safety culture site to site.

ASSESSING WHETHER GOALS HAVE BEEN MET

This lead us into the topic of determining if safety goals have been met. There is a necessary distinction between such an assessment for a specific plant, and assessment for an entire industry. The former has just been discussed in connection with the use of PSA to meet safety goals.

Assessment of the safety of an entire industry can in principle be done in two ways, through PSA estimates for all plants in the industry, or through historical evidence. Each has its problems.

The estimate based on probabilistic safety assessment would be obtained by averaging probabilities for individual plants. For instance, if the safety goal is based on the probability of severe core damage, the assessment would simply be in terms of the average value of this parameter for the plants. The problem is that the uncertainty margins of the values for individual plants are large. The uncertainty of the average is improved to some extent by the averaging process, but not as much as would be hoped. The data bases for all of the plants have much in common, and this tends to propagate uncertainties through the analysis as systematic errors. Furthermore, the uncertainty margins of the few plants with higher contributions to the averaging process tend to dominate. Yet it should be expected that some benefit is gained from averaging. Of course, another difficulty is that the use of PSA in this way requires that PSA's exist for all of the plants.

The historical method provides an estimate from the record of frequency of accidents. This method is really only fruitful for water reactors, because it demands a large number of operating years of experience.

The approximately 350 water reactors in the world have now accumulated about 5,000 reactor years of operation. One instance of severe core damage has occurred, at Three Mile Island. Experience is being gathered at a rate of about 1,000 reactor years every three years, and in fifteen years the record will have become one case of severe core damage every 10^4 years (assuming that no further cases occur). This will be at a level of a number of safety goals for existing reactors.

Since no fatalities have occurred from accidents to plants with water reactors, and the number of estimated cancers resulting from the accident to TMI is less than one, the record now meets the safety goals in the United States, which are stated in terms of the consequences of accidents.

As safety goals are set at smaller and smaller probabilities of undesired effects, the ability to assess their achievement by any means becomes more difficult. The INSAG goal on safety of future plants is a probability of severe damage of less than 10^{-5} per reactor year. This value has also been adopted by a number of other groups. But some countries have adopted a goal at a probability of 10^{-6} per reactor year. The difficulties in assessment are evident.

I have been able to think of only one method to help in such a situation. PSA can be applied to estimate the rate at which situations occur that if uninterrupted could lead to severe core damage. This may have a rate of occurrence that can be observed historically. Precursor studies are of this form, but accident precursors are too near the onset of accident sequences and do not reflect the ability of the plant to survive. What is needed instead is something like a historical record that does test the design of the plant and its operating practices.

DISCUSSIONS

F. COGNE, France

J'ouvre maintenant les discussions sur les communications présentées ce matin.

J. CHANTEUR, France

M. Silini, vous n'avez fait aucune allusion à l'hormesis. Comme les désordres multifactoriels, cet effet ne peut effectivement être quantifié avec certitude. Mais pourquoi parler des uns (les désordres) sans évoquer l'autre ?

Vous avez indiqué que l'on n'avait aucune expérience humaine directe de cancer induit. C'est vrai à faible dose, mais ce ne l'est pas à dose élevée où l'on connaît bien, notamment, les cancers secondaires à la radiothérapie.

Pourquoi, en matière d'évaluation du risque radiologique, réduire l'exposé au cas des faibles doses, ce qui finit par suggérer qu'elles sont plus redoutables que les fortes ?

Vous avez en effet développé plus largement l'exposé des effets stochastiques en déclarant : "nous pensons que la probabilité d'induction de ces effets est augmentée, même pour les plus petites doses". C'est une sorte de paralogisme. Pourquoi le pensons-nous ? Parce que nous en sommes persuadés. "Et voilà pourquoi votre femme est muette" ! Ne sommes-nous pas entraînés dans une axiomatique détachée de l'expérience et fondée sur une conviction qui tend actuellement à s'imposer comme une idéologie dominante hors de laquelle il est malséant de se situer ? Permettez-moi de rappeler que tous les experts ne sont pas convaincus de la validité de cette axiomatique surréaliste. En tout cas, contrairement à ce que je viens d'entendre, la relation linéaire permet de prévenir les effets stochastiques, pas de les prévoir.

G. SILINI, Italy

The existence of multifactorial conditions is well known and accepted. Hormesis is a theory which, as far as I know, has not been convincingly demonstrated.

The question of secondary cancer is well dealt with in the 1988 UNSCEAR report. To use these data in order to quantify cancer radiation risk is at present not possible because one does not know to what extent the people who have developed cancer before are representative of the total population.

Your third question is centered around the question of the non-threshold dose linearity. I believe that this question may be raised also during the rest of the symposium. I should like, therefore, to state the reasons why I believe it should be applied.

a) For scientific reasons, first. When the dose and dose rate are low and effects depend on the traversal of few cells by single tracks, the induction of primary damage must be linear. Repair phenomena, which are often cited to oppose this argument, are not valid because as long as repair itself is not dose-dependent (which could only occur at high doses), repair could only decrease the probability of an effect but not its linearity. In any case, the risk estimates are based on tumours that have actually been observed and for which any primary and repair phenomena have already taken place.

b) For observational reasons. In Hiroshima and Nagasaki, as one approaches lower and lower doses, one finds, at least for all solid tumours, dose-response relationships that are nearly linear.

c) For practical reasons, because we are thinking of doses above a fairly high background. For small increments above it a linear approximation will always be acceptable. Furthermore, if we should postulate anything else than non-threshold linearity, we would be forced to keep track of each dose received by each person to calculate the level of risk corresponding to each increment. This would be almost impossible. Thus we should recur to some form of averaging, which would be equivalent to assuming linearity in the first place.

In conclusion, I believe that many things will be likely to change in radiation protection; but the hypothesis of non-threshold linearity will stay with us for the time being.

D. BENINSON, ICRP

Among the assumptions you quoted regarding hereditary effects, Dr. Silini, there is one which is not an assumption: the equal value per generation at equilibrium and the total for a single irradiation; this is just a mathematical consequence.

G. SILINI, Italy

It is true that the number of hereditary effects at equilibrium may be shown mathematically to be the same as the number per single irradiation, with a procedure analogous to that used for calculation of dose commitment. But mathematics should be proved by biology, which may be impossible in man.

A.P.U. VUORINEN, Finland

How strong is the evidence that cancer induction of ionizing radiation is stochastic in its character? The other possibility for stochastic appearance could be statistic character of sensitivity for cancer among individuals.

G. SILINI, Italy

For the purpose of radiation protection, the definition of stochastic effects is meant to apply at the descriptive clinical level; it is fair enough to say it is stochastic. It is well known that, at least for some human types, susceptibility to some forms of cancer is higher in some groups of people, who are a small percentage of the total population. Therefore, one would never expose to radiation people known to be affected by ataxia teleangectasy or Fanconi's anaemia. Radiation protection recommendations are based on the concept of an average susceptibility and not tailored for single individuals. The extreme cases that you are alluding to must be handled apart, within the context of occupational medicine.

S. BENASSAI, Italy

This question relates to the consequences of the Chernobyl accident. Are there some available data concerning malformations or other radiation-induced effects concerning the irradiation in utero of embryos and fetus?

G. SILINI, Italy

Apart from undocumented press reports, I am not aware of any good scientific publication on this subject.

A.J. GONZALEZ, IAEA

As you implicitly indicated in your paper, Dr. Silini, the necessary melting of a convincing radiobiological theory and the correlated observational evidence has been achieved for radiocarcinogenic effects but not yet for radio-hereditary effects (at least in human beings). In spite of this, however, risk estimates have been made for both types of effects, seemingly without conceptual differentiation. Some nuclear safety specialists have seen this approach as not necessarily coherent and consistent. Could you comment on this?

G. SILINI, Italy

I have stated explicitly in my paper what are the sources of information available to assess the risk in the somatic and genetic field. The duty of the assessors is to provide the best possible estimates and to indicate their limitations and weaknessess. But it would be unfair to attribute to the evaluators' lack of coherence and consistency what is simply a lack of primary information.

H. KOUTS, United States

I would like to emphasize the thought implied by Dr. Silini and stated by Dr. Clarke, that there is an element called prudence in the ICRP's recommendations. Speaking for the nuclear safety community, I would like to see the term "prudence" retained in the statements of ICRP policy.

R. H. CLARKE, United Kingdom

Our objective is not to underestimate the risk. We don't want to be in a position in a few years time to have to increase risk estimates once more. Moreover, human data on hereditary risk do not contradict the risk figures we are using at the moment, and, secondly, if we take the combined worker epidemiology studies, the overall cancer risk to date is consistent with the risk estimate we are now making.

D. BENINSON, ICRP

On the word "prudence". We are here in a situation of defining terms. We are reluctant to use the word "prudent" because it was used before to mean an assumption of linearity that was derived directly by extrapolation from observed data at high doses down to zero. That gave an overestimate in risk estimates for most parts of the curve and was then taken to be prudent. This is not done any longer. There is much more use of a linear quadratic relationship which is based on cellular radiobiology and, therefore, the present assessments are more realistic. This doesn't mean that they are not uncertain, but they are not meant to be on the conservative side. They are meant to be as realistic as possible.

R.J. BERRY, United Kingdom

Can we please clarify the scientific basis of human risk estimates: for all the occupationally-exposed populations from the nuclear industry (not the early radiologists who were subjected to deterministic damage) there is so far no evidence of an excess of cancers. What has been shown is a trend for an increase in cancer incidence for particular cancers which is related to cumulative occupational radiation exposure – but in no case is the total number of cancers in any site in excess of the expectation in a normal population. Thus, there is still a bottom line compatible with observation in humans that at current levels of occupational exposure there is no excess incidence of malignant disease.

G. SILINI, Italy

Obviously, due to the long latency times, one would have to wait for a few decades in order to see any cancer due to current levels of occupational exposure. But, in any case, the limits for planned exposures are set well below the level of detectability for the current spontaneous incidence.

D. BENINSON, ICRP

When we discuss the consequences of the risk factor in terms of cancer cases, we have to realise the statistics involved. Even if our estimates are meant to be realistic we risk not being able to see the effects of them in the population Dr. Berry talked about. We do not have sufficient "power" in the statistics. Therefore, it is not surprising that we do not see any effects, even if the effects were quite real. Concerning genetics, it doesn't surprise me that we do not see any evidence yet for genetic damage in man, while we see it in other species. The generation time in man is quite long, compared to some animal species.

P.Y. TANGUY, France

Je partage le point de vue exprimé par le Dr. Kouts à la fin de sa présentation : une validation sur les incidents observés en exploitation permettrait d'accroître la confiance en ce qui concerne les évaluations probabilistes. Ma question porte sur la démonstration d'objectifs de l'ordre de 10^{-6} par an pour la dégradation du coeur. Elle me paraîtrait plus aisée si d'une part, la part de l'intervention humaine était minimisée dans le risque, et d'autre part, les installations étaient plus simples, afin d'être transparentes pour les analystes de sûreté comme pour les exploitants.

H. KOUTS, United States

I thoroughly agree with Mr. Tanguy. To carry probabilities to these low values it will be necessary to reduce the impact of personal actions on risk, and to simplify plants. Then perhaps the PSA would be a much more reliable guide to whether or not you can achieve your goals.

L.G. HÖGBERG, Sweden

I agree with what Dr. Kouts said about improving precision of PSA risk estimates through more automated plants, leaving less room for human errors; however, I would like to point out some limiting conditions for such improvements:

(1) You must improve simplicity of design and transparency of design, otherwise uncertainties associated with the completeness of PSA remain or may even increase due to difficulties of finding all "subtle" human errors in the design of complex systems and associated testing and maintenance activities.

(2) You must design reactor systems characterised by simplicity in diagnosis and management of abnormal conditions which are outside the design basis of the automated systems or in which the automated systems malfunction. For example, even state-of-art twin-computer systems typically fail for several hours per year, often correlated with plant disturbances creating a high signal load. In such cases you have to rely on human intervention to manage the situation under conditions that may be sensitive to errors of commission, which are difficult to estimate with current PSA methods.

H. KOUTS, United States

You have made an important point. There certainly are limits to reduction of the effect of human factors.

E.A. RYDER, United Kingdom

I wish to return to the point discussed between Messrs Kouts, Tanguy and Mr. Högberg as to whether nuclear safety is better served by reliance on operators or engineered systems. I understand the points made very well. I merely wish to comment that there have been two spectacular accidents at nuclear power plants; one at Three Mile Island, the other at Chernobyl; both were operator-assisted. It is therefore very easy to understand the strong trend to avoid too much reliance on the operators. Whether this is best done by adding safety systems to the well understood and developed existing designs or whether by changing to advanced designs, with greater reliance on passive features, remains to be seen.

D. BENINSON, ICRP

Even with old experimental psychology, human reliability is insufficient compared with other components of the "system", supporting strongly the concept of less need of human action for better safety.

H. KOUTS, United States

Reduction of the role of human beings is important to reducing risk and uncertainty in risk, but one must remember Högberg's remark that the effect of human fallibility will always remain to some extent.

Session 2

BASIC OBJECTIVES AND POLICIES

Séance 2

OBJECTIFS ET POLITIQUES : ASPECTS FONDAMENTAUX

Chairman – Président

**R.H. CLARKE
(United Kingdom)**

LA DOCTRINE DE LA CIPR EN MATIERE DE RADIOPROTECTION ET SON APPLICATION

D. Beninson
Comisión Nacional de Energía Atómica
Buenos Aires, Argentine

RESUME

La doctrine de la CIPR en matière de radioprotection est décrite dans les principales Recommandations qu'elle publie périodiquement. La présente communication contient un examen de certains des aspects de cette doctrine, qui se fonde sur les nouvelles Recommandations dont l'adoption est attendue d'ici à la fin de l'année. Dans ces Recommandations, la Commission ne modifiera pas les principes fondamentaux de son système de radioprotection, à savoir que les pratiques à l'origine d'expositions devraient être justifiées, que les dispositions en vue d'assurer la protection devraient être optimisées, et que les expositions individuelles devraient être restreintes, selon le cas, par des limites de dose, des contraintes de dose liées à la source ou des limites prescrites. Le fait que la Commission établira, dans ses recommandations, une nette différenciation entre les pratiques donnant lieu à des radioexpositions, lesquelles exigent l'application du système de protection, et les situations dans lesquelles les expositions existantes aux rayonnements nécessitent des décisions visant des actions correctives, constitue un important progrès sur le plan conceptuel. Dans son examen de la radiobiologie et de l'épidémiologie, la Commission conclut que les nouvelles données et une nouvelle interprétation des informations antérieures montrent maintenant avec une certitude raisonnable que les risques liés aux rayonnements ionisants sont environ trois fois plus élevés qu'on ne l'estimait il y a une dizaine d'années. Cette augmentation impose d'apporter certaines modifications quantitatives aux recommandations de la Commision. L'une de ces modifications, qui sera recommandée par la Commission, est une réduction de la limite de dose applicable à la radioexposition professionnelle. Le chiffre actuel de 50 millisieverts par an sera ramené à 20 millisieverts par an, certaines dispositions étant prévues pour ménager de la souplesse d'une année sur l'autre. L'actuelle limite relative à la dose à long terme pour les personnes du public (1 millisievert par an en moyenne sur la durée de vie) sera conservée, mais une protection accrue sera assurée du fait que la période retenue pour la moyenne sera limitée à cinq ans et que le recours à des contraintes supplémentaires liées à la source sera recommandé.

THE ICRP RADIATION PROTECTION PHILOSOPHY AND ITS APPLICATIONS

D. Beninson
Comisión Nacional de Energía Atómica
Buenos Aires, Argentina

ABSTRACT

The ICRP philosophy of radiation protection is described in the main Recommendations issued periodically. The paper reviews some issues of this philosophy on the bases of the new Recommendations expected to be approved by the end of the year. In these Recommendations, the Commission will not change the basic principles of its system of radiation protection, namely that practices causing exposures should be justified, the protection arrangements should be optimized, and the individual exposures should be restricted by dose limits, source-related dose constraints or prescribed limits, as appropriate. As an important conceptual development, the Commission will clearly differentiate in its recommendations between practices giving rise to radiation exposures, calling for the application of the system of protection, and situations where existing radiation exposures require decisions on remedial actions. In its review of radiobiology and epidemiology, the Commission concludes that new data and new interpretation of earlier information now indicate with reasonable certainty that the risks associated with ionizing radiation are about three times higher than they were estimated to be a decade ago. This increase calls for some quantitative changes in the Commission's recommendations. One such change to be recommended by the Commission is a reduction of the dose limit for occupational exposure. The current figure of 50 millisievert in a year will be reduced to 20 millisievert in a year, with some provisions to allow year-to-year flexibility. The current limit for the long-term dose for the public (1 millisievert per year averaged over a lifetime) will be retained, but increased protection will be provided by limiting the averaging period to five years and recommending the use of additional source-related constraints.

1   INTRODUCTION

It is now expected that the new main ICRP Recommendations will be approved next November, after a substantial period of interactive elaboration, dissemination for comments to each time wider audiences, reviewing and consolidation.

The new Recommendations can be described as the result of an evolutionary process rather than being revolutionary, when compared to the previous Recommendations published in 1977(1). During the intervening period those Recommendations were amended by a Statement in 1978 and further clarified and extended by Statements in 1980, 1983, 1984, 1985, 1987 and 1989(2). Furthermore, new data and new interpretations of earlier information indicate with reasonable certainty that the risks per unit dose associated with ionizing radiation are higher than they were estimated to be a decade ago.

The present paper summarises some of the main issues of the new Recommendations, with the addition of comments and opinions of the author. While the author is deeply involved with the work of the ICRP, such comments and opinions should not be taken to represent necessarily the views of the ICRP.

2.   SCOPE OF THE RECOMMENDATIONS

As previously, the new Recommendations are confined to protection of man against ionising radiation. While dealing with only one of the many dangers facing man, they emphasise the view that ionising radiation should be treated with care rather than fear and that, for assessments and decision-making, its risks should be kept in perspective with other risks.

The main aim of the Recommendations is to provide an adequate standard of protection for man without unduly limiting the beneficial practices giving rise to radiation exposures. The Recommendations do not deal with other species than man, but it seems most likely that the degree of environmental control needed to protect man at the present level of ambition will ensure that other species are not put at risk, at least as populations or species. For this reason, the environment is only considered regarding the transfer of radionuclides through it, affecting the protection of man.

3.   QUANTITIES USED IN RADIATION PROTECTION

It has always been apparent that the sophistication involved in the definition and specification of radiation protection quantities was beyond balance with the radiobiological knowledge, its uncertainties and with the practice of protection.

The fundamental dosimetric quantity is the absorbed dose. It is defined as a point quantity, but in radiation protection is used as the average over an organ or tissue. The use of the average dose as an indicator of the probability of inducing stochastic effects rests on the linear non-threshold relationship described in section 4, which is a reasonable approximation over the limited range of dose relevant in practice. At higher doses, where deterministic effects occurr, the average dose is relevant to such deterministic effect only when dose is fairly uniform over the organ or tissue.

Dose equivalent, absorbed dose times the quality factor Q (function of LET), is also a point quantity. The new Recommendations continue to use dose equivalent for measurments of operational quantities such as "ambient dose equivalent".

However, in radiation protection, a weighted average absorbed dose over organs or tissues is more relevant in terms of health effects. Instead of using any function of LET, direct RBE can be determined experimentally for different radiation incident upon the body or from sources within the body.

The new Recommendations define, for protection purposes, a quantity <u>equivalent dose</u>, H_T , as

$$H_T = w_R \, D_T$$

where D_T is the average absorbed dose in the organ or tissue T and w_R is a weighting factor selected for the type and energy of the radiation incident on the body or emitted by a source within the body.

The values of w_R proposed by the new Recommendations are stylised numbers base on the review of RBE included in the Biolo<u>gical</u> Annex to the main recommendations.

Radiation type and energy	w_R
Photons, all energies	1
Electrons and muons (all energies)	1
Neutrons < 10 keV	5
> 10 keV - 100 keV	10
> 100 keV - 2 MeV	20
> 2 MeV - 20 MeV	10
> 20 MeV	5
Protons > 20 MeV	5
Alpha particles, fission fragments and heavy nuclei	20

For convenience an approximation obtained by fitting to the

w_R for neutrons can be expressed as a continuous function of incident energy. For cases where the w_R value is not specified, it would be possible to obtain an approximate value by the use of the Q-LET relationship.

The relationship between the probability of stochastic effects and equivalent dose also depends on the radiosensitivity of the irradiated organ or tissue. In order to provide a single quantity relating to the stochastic risk, the Recommendation, as the previous one, use the concept of Effective dose, E ,defined as

$$E = \sum_R w_R \sum_T w_T D_{T,R} = \sum_T w_T \sum_R w_R D_{T,R}$$

where $D_{T,R}$ is the average absorbed dose in organ or tissue T delivered by radiation R and w_T is the weighting factor for tissue or organ T.

The values of w_T have been developed taking into account somatic stochastic harm (lethal malignancies and non-lethal ones) and hereditary effects. The probability of non-lethal cancers is weighted by their "lethality fraction", as it was suggested in ICRP Publication 45. The new Recommendations contain a substantial discussion of the w_T , but for practical reasons and in view of the uncertainties, the values are grouped as follows:

w_T	0.01	0.05	0.12	0.20
Organs or Tissues	Skin Bone Surface	Thyroid Liver Orsophagus Breast Bladder "Remainder"	Lung Red bone marrow Stomach Colon	Gonads

The "remainder" w_T applies to the average dose in the unlisted part of the body (total energy deposited divided by total mass). However, the Recommendations state that when an organ or tissue without a specific w_T receives substantial doses, half of the w_T for the "remainder" be allocated to that organ or tissue.

4. RADIOBIOLOGICAL BASES OF RADIATION PROTECTION

Deterministic effects are the result of unviable changes in

many cells of given organ or tissues. Such changes in individual cells are of random nature, but the required large number of cells makes the effect deterministic, with thresholds. It is therefore possible to prevent the occurrence of deterministic effects by keeping doses below the relevant thresholds.

As in the 1977 Recommendations, the basis for the Commission new Recommendations is the linear non-threshold relation--ship between probability of induction of stochastic effects and dose. As epidemiological data will never be sufficient to determine the shape of the relationship at low doses, any extrapolation from observations at the higher doses must be based on models of sufficient "descriptive realism".

The model used by the Commission is briefly described in the following paragraphs. The occurrence of ionization and excitation events, and of cell damage, is a random process and the number of damage events at a given dose follows a Poissonian distribution, the dose being proportional to the average of such random variable. If cellular damage doses occur, it may be repaired by the cell, it may prevent the survival or reproduction of the cell, or it may result in a viable but transformed cell.

There is increasing evidence that cellular DNA is the principal target for radiation-induced sterilization of cells and the induction of chromosomal and other genetic changes. In order to deal with the initial DNA damage that gives rise to these changes, cells have evolved complex, enzyme-mediated repair systems. These are often specific for different molecular forms of DNA damage, and identify and remove the lesions induced in DNA by ionizing radiation, and also by ultra-violet and chemical agents.

Such induced damage may be repaired with high fidelity, returning the DNA structure to its original form (error-free repair) in which case there is no long-term cellular consequence of that lesion. Alternatively, with smaller probability repair processes may act on a more error-prone manner that retains over all DNA integrity but results in small base sequence changes (point mutations) at the site of the initial lesions or more gross changes such as gene deletions or rearrangements. These misrepair events. if they occur in important regions of DNA, may have long term consequences for the cell and can result in cell sterilization or stable genetic changes in surviving cells.

There is strong evidence that misrepair becomes more frequent in the case of double strand breaks of DNA, which require two damage events. The probability of such two events, occurring within a short period depend on the square of the dose and, of course, on dose rate.

While single cell death is irrelevant, a single transformed but viable cell may have far reaching consequences. A transformed _somatic_ cell may be able to multiply to form a developing clone of modified cells that may eventually give rise to a malig nant tumor, a cancer. A transformed _germ_ cell doses not form a clone but can transmit incorrect hereditary data to the descend- ants of an irradiated individual. Both these effects, which may start from a single modified cell, are known as "stochastic" effects.

The response of the body to the development of a clone of modified somatic cells is complex. Such a clone is almost always eliminated or isolated by the body's defenses, but, if it is not, it may result, after a prolonged delay called the latent period, in the development of a malignant condition in which the prolif- eration of cells is uncontrolled. The defense mechanisms are not totally effective, so the probability of a cancer resulting from the radiation is at least partly dependent on the number of clones of modified cells initially transformed, since this num- ber will influence the probability of at least one clone sur- viving.

It is then the probability of malignancy, that is related to dose, while the severity is influenced only by the type and location of the malignant condition. The process appears to be random, although it is likely that individuals differ in their sensitivities, to the induction of cancer by radiation. It seems that no somatic stochastic injuries other than cancer are in- duced by low and moderate doses of radiation.

The simplest relationship between dose equivalent and the probability of a defined stochastic effect is that of a straight line through the origin, at the low doses. This is the type of relationship that is expected from microdosimetric considera- tions, which predict a linear relationship whenever there is, on average, less than one particle track per cell, namely at low doses. As single events are involved, dose rate is not expected to influence the relationship at these low doses where linearity applies.

Almost all the data relating to stochastic changes in cells "in vitro", and to many animal tumor systems show curvilinear dose-effect relationships for radiations of low linear energy transfer (LET), with a linear component at low doses with a smaller slope than that at high doses. The results for radiation of high LET are usually more nearly rectilinear, with some show- ing even an increased slope at low doses. A complication is in- troduced by the effect of cell killing at high absorbed doses (1 Gy or more). This causes a reduction in the slope of the dose effect relationship for stochastic effects at high doses, partic ularly at high dose rates. Data on mamalian cancers and mutations are broadly consistent with the data on cells, but the statisti- cal limitations in these studies preclude definite confirmation. The human data are not sufficiently precise to confirm or exclude that relationship. The data for many types of cancer suggest this form; the data for some others do not.

In conclusion, the most characteristic form of the relationship between the dose equivalent in an organ and the probability of a resultant cancer is that of an initial proportional response at low values of dose equivalent, followed by a steeper rate of increase (slope) that can be represented by a quadratic term, followed finally by a decreasing slope due to cell killing There are no adequate grounds for assuming a threshold in the relationship. However, the slope of the linear component (due to misrepaired single damage events in DNA) is smaller than the slope of extrapolations to the origin of statistical data obtained at high doses (and dose rates), where the quadratic component is substantial, because of the more common misrepaired double strand damage.

Human epidemiological data are available at high doses. The derivation of the slope of the linear component, relevant for radiation protection, must rely on the "descriptive realistic" model and uses the so-called "dose and dose rate factor", which is known with some uncertainty.

Regarding the risk estimates for fatal cancers, the Commission believes that new data now indicate with reasonable certainty that the risks associated with ionising radiation are about three times higher than they were estimated to be a decade ago. The UNSCEAR (1988) estimate of probability for lifetime fatal cancers using the preferred multiplicative risk projection model is 11×10^{-2} Sv^{-1} for the exposed populations at Hiroshima and Nagasaki, of whom two thirds in the epidemiological study are still alive. Estimates of the probability of radiation-induced lifetime fatal cancers in five populations, including the Japanese, using two different multiplicative projection models and the base line cancer mortality rate appropriate for each country, yield an average value of about 10×10^{-2} Sv^{-1}.

Applying an effectiveness factor of 2 to allow for repair processes at low doses and low dose rates yields a value of 5×10^{-2} Sv^{-1} for the probability of lifetime radiation-induced fatal cancers in a nominal population of all ages. A smaller average value at about 4×10^{-2} Sv^{-1} would be obtained for a working population aged 20 - 64 years, exposed during their working life.

A justification of the primary use of Japanese epidemiological data, the dose-effect relationship, the risk projection models, dose and dose rate effectiveness factor, and details of the risks for site-specific cancers will be given in the Biological annex of the new Recommendations.

The new Recommendations discuss also heredity effects and mental retardation from in utero irradiation, and a scientific review of the available information is included in the Radiobiological Annex to the Recommendations. The following table summarises the nominal coefficients used in the new Recommendations

Biological effect	Exposed population	Risk coefficient $(10^{-2}\ Sv^{-1})$
Fatal cancer	Adult workers	4.0
Fatal cancer	Whole population	5.0
Serious hereditary effects	Adult workers	0.6
	Whole population	1.0
Mental retardation	8-15 week conceptus	30 IQ points Sv^{-1}

It should be noted that mental retardation is not really a stochastic effect; the value given under risk coefficient is really a linear shift of the IQ distribution curve.

5. THE CONCEPT OF DETRIMENT

As in the past, the new Recommendations find useful to take account of the different probabilities and severities of the effects. This is mainly done for the assignment of values to the tissue weighting factors w_T. On the other hand, such "distillation" of probabilities and severities of several possible stochastic effects does not describe realistically the exposure situation.

For this reason, the new Recommendations replace the aggregative concept of detriment by a multiattribute concept for comparing the impact of radiation with that of other hazards. For example, the attributes associated with mortality considered are: the lifetime attributable probability of death, the life lost if the attributable death occurs, the reduction of life expectancy, the time distribution of the attributable probability of death and the increase in the "mortality force".

Morbidity due to non-fatal cancers and hereditary effects are taken into account by weighting for severity and for life lost and impaired.

Comparisons of the consequence of given exposure situations with that of other hazards are carried out in the Annex C to the new Recommendations, as a basis for selecting dose limits.

6. THE CONCEPTUAL FRAMEWORK OF RADIATION PROTECTION

Everyone in the world is exposed to radiation from natural and artificial sources. Any realistic system of radiological protection must therefore have a clearly defined scope if it is not to apply to the whole of mankind's activities. It also has to cover, in a consistent way, a very wide range of circumstances. For many purposes, each source of exposure can be treated on its own. Each individual, however, is exposed as a result of several sources. It follows that assessments of the effectiveness of protection can be related to the source giving rise to the individual doses (source-related) or related to the individual dose received by a person from all the relevant sources (individual-related).

Some human activities increase the overall exposure to radiation, either by introducing new sources of exposure, or by increasing the number of individuals exposed. The Commission calls these human activities "practices". Other human activities can decrease the overall exposure by influencing the existing exposure pathways between the sources and man. The Commission descirbes these activities as "intervention".

The linear non-threshold relationship for the induction of stochastic effects implies that a given increment of dose causes the same increment of risk irrespective of previously accumulated dose or of future doses that might be incurred. It is therefore possible to stipulate "constraints" for the dose from any practice or defined combination of practices.

Under the constraints the risks are still not zero. The following considerations apply:

a) Any level of dose implies a probability of stochastic effects (which may be fatal malignancies and severe genetic effects). Therefore, at any level of protection applied to the source, there is an expectation of some harm caused by the source.

b) The radiation detriment due to a source is a function of the distribution of all the radiation exposures, either present or future, caused by the source during its operating lifetime. Under the proportionality relationship the expected number of attributable stochastic effects is proportional to the total collective effective dose from the source.

c) Any level of protection applied to the source implies a social cost which, by diverting resources from beneficial uses, can be considered also as detrimental to society.

d)   Conceptually, therefore, there should be a level of
protection for a given source which minimizes the com-
bines detriment to society resulting from the cost of
protection and the radiation detriment from the source.

These consideratons underline the requirement of "optimization of protection", namely to keep doses "as low as reasonably
achievable, taking into account social and economical considerations". This requirement is paramount in the case of exposures
from practices.

The other two components of the system of radiation protection, justification of the practice and individual dose limits,
also continue to be recommended. The new Recommendations include
in the protection for practices the control of risk in the case
of potential exposures. A conceptual example is presente below.

Exposures might occur as a result of disruptive events. The
risk in this case depends both on the probability of the event
and on the resulting radiation dose. Regarding radiation effects,
one deals, therefore, with effects of second order stochasticity.
For example, the probability of dying due to a given potential
disruptive event can be expressed as $R = P_1 P_2$, where R is
the nominal risk, P_1 is the probability of occurrence of the
event which would result in a dose to the individual under consideration, and P_2 is the conditional probability of death given
that dose. It should be realized that P_2 is proportional to
dose at the lower doses (where only stochastic effects are possible), but increases steeply with whole-body dose at values
above a few gray due to acute non-stochastic lethal effects.

If several potential disruptive events are taken to be possible, the nominal risk R given in the previous paragraph becomes:

$$R = \sum_i P_{1i} P_{2i}$$

It must be noted, however, that the nominal risk as defined
above for potential exposures from disruptive events, does not
cover all harmful effects which are possible at the higher doses,
but is a possible quantity for expressions where the attributable
mortality probability is relevent.

Several misconceptions exist regarding the function of dose
(risk) limits. Firstly, the dose limit is widely, but erroneously, regarded as a line of demarcation between "safe" and "danger
ous". Secondly it is also erroneously regarded as the only effective way of controlling exposures. Thirdly it is seen, also
wrongly, as the sole measure of the stringency of a system of
radiation protection.

The conceptual change of "detriment" and the ways of judging it, and the new assessments of the radiation risks, call for some quantitative changes in the Commission's recommendations. One such change to be recommended by the Commission is a reduction of the dose limit for occupational exposure. The current figure of 50 millisievert in a year will be reduced to 20 millisievert in a year, with some provisions to allow year-to--year flexibility. The current limit for the long-term dose for the public (1 millisievert per year averaged over a lifetime) will be retained, but increased protection will be provided by limiting the averaging period to five years and recommending the use of additional source-related constraints.

When the sources of exposure and the exposure pathways are already present, the only type of action available is <u>intervention</u>. Before any intervention is iniciated it should be monstrated that it will do more good than harm (<u>justification of intervention</u>). It should be stressed that the <u>intervention "cost"</u> is not just the monetary cost: some remedial actions may involve non-radiological risks or serious social impacts.

The method, scale, radiation dose at which to introduce the action and duration of intervention should be optimised so as to obtain the maximum net benefit. Dose limits are, of course, irrelevant for remedial situations.

7.   SOME ISSUES IN IMPLEMENTATION

In previous recommendations, the Commission has defined two types of working conditions based on the likehood of reaching a level of individual annual dose. This was originally intended to help in the choice of workers to be subject to individual monitoring and special medical surveillance. In recent years, it has become apparent that working conditions are best based on expected dose and the Commission no longer recommends such a classification.

The control of sources by management is helped by requiring that the workplaces containing them be formally designated. The Commission uses two such designations - controlled areas and supervised areas. A controlled area is one in which normal working conditions, including the possible occurrence of minor mishaps, require the workers to follow well established procedures and practices aimed specifically at controlling radiation exposures. A supervised area is one in which the working conditions are kept under review but special procedures are not normally needed.

The definitions are best based on operational experience and judgement. The Commission now regards the 3/10th concept dividing between controlled and supervised areas as being too arbitrary and recommends that the designation of controlled and supervised areas should be decided either at the design stage or locally by the operating management on the basis of operational experience and judgement.

In order to avoid excessive regulatory procedures, most regulatory systems include provisions for granting exemptions in cases where it is clear that a practice is justified, but where regulatory provisions are unnecessary.

The Commission notes that the IAEA and NEA have issued advice on the subject and believes that exemption of sources is an important component of regulatory functions.

Sources that are essentially uncontrollable, such as cosmic radiation at ground level and potassium-40 in the body, can best be dealt with by the process of exclusion from the scope of the regulatory instruments, rather than by an exemption provision forming part of the regulatory instruments.

References

(1)   ICRP Publication 26. Recommendations of the ICRP. Annals of the ICRP 1 (3), 1977.

(2)   Statements in the Annals of the ICRP

Statement from the 1978 Stockholm Meeting of the ICRP 2 (1) 1978.

Statement and Recommendations of the 1980 Brighton Meeting of the ICRP 4 (3/4) 1980.

Statement from the 1983 Washington Meeting of the ICRP 14, i-vii, 1984.

Statement from the 1984 Stockholm Meeting of the ICRP 14 (2), 1984.

Statement from the 1985 Paris Meeting of the ICRP 15 (3), i-ii, 1985.

Statement from the 1987 Washington Meeting of the ICRP 17 (2/3), i-iii, 1987.

Statement from the 1987 Como Meeting of the ICRP 17 (4), i-v, 1987.

Statement from the 1989 Paris Meeting of the ICRP 20, (1), 1989.

DISCUSSIONS

R.H. CLARKE, United Kingdom

Questions to Dr. Beninson, please.

O. ILARI, NEA

Dr. Beninson, you mentioned a few reasons why the concept of dose limit is wrongly interpreted and used by many. One is that the level of dose limit is wrongly taken as an indicator of the stringency of the radiation protection requirement. In effect, you say, one can have a very good level of protection even with a high value of the dose limit, because optimisation is the main tool to improve protection.

I agree with all this, of course, but in this case many persons in this audience could ask why the ICRP feels a need to reduce the dose limit for workers against the opinion of those who were worried that such a reduction would narrow the margin available for putting the optimisation tool into play.

D. BENINSON, ICRP

In many cases optimisation is the more stringent requirement. For example, in some plants shielding has been designed (by optimisation) to correspond to a maximum annual effective dose of 5 mSv. In these cases, the limit is a bad indicator (50 mSv or 20 mSv would not influence the result). In other cases, however, optimisation would not reduce doses much below the limit (for example for workers in underground uranium mines). In these cases the limit is important and stringent. As the ICRP Recommendations apply to all practices, they must contain advice about the dose limit, even if in most cases (if optimisation is really used), such a dose limit is a bad indicator of the goodness of a radiation protection programme.

R.J. BERRY, United Kingdom

In view of your comments on the misuse of dose limits, how can you defend the "magical" importance of specifying these over one year. Would it not be more logical to have a longer period, perhaps even a lifetime limit which would then allow both flexibility of employment and limitation of lifetime risk - which should be the aim?

D. BENINSON, ICRP

Lifetime limits would have the disadvantage of permitting radiation safety relaxation today on the expectation of improving situations in the future or because good safety existed in the past. On the other hand, the new ICRP Recommendations will provide flexibility, giving a period of averaging of annual doses.

A.P.U. VUORINEN, Finland

When talking about dose limits you seem to be very sure that you are right and it might be so when looking at radiation protection philosophy from your special, rather narrow angle.

I think that other aspects have to be considered, e.g. in relation to decision making in our society and at the level of practical work at nuclear power plants. Even though your approach strictly taken is logical and provides a sound rational basis for thinking, it might in some situations be unpractical and lead to unnecessary, unexpected losses in the society.

D. BENINSON, ICRP

The presentation was based on science and meant to be a consistent approach; I think this is the advice required by decision-makers. They would take account of the other factors (which could be considered more political). If the scientist already include these political issues in his advice, political issues would be counted twice.

THE UNITED STATES SAFETY GOALS

F.J. Remick
U.S. Nuclear Regulatory Commission
Washington, D.C., United States

ABSTRACT

Following the accident at Three Mile Island, the U.S. Nuclear
Regulatory Commission imposed many new requirements. The NRC's
credibility with the general public was at an all time low, while
regulatory instability was at an all time high. In partial
response to those concerns, the NRC issued in 1986 a policy
statement on safety goals. It became obvious that implementation
of the safety goals was not an easy task. This paper describes
recent Commission directives to its staff on implementation of
the Safety Goals Policy Statement.

LES OBJECTIFS DE SURETE ADOPTES AUX ETATS-UNIS

RESUME

A la suite de l'accident survenu à Three Mile Island, la Commission de
la réglementation nucléaire (NRC) des Etats-Unis a imposé de nombreuses pres-
criptions nouvelles. La crédibilité de la NRC auprès du grand public était
tombée à son niveau le plus bas, alors que l'instabilité réglementaire battait
tous les records. En partie en réponse à ces préoccupations, la NRC a diffusé
en 1986 une déclaration d'orientations sur les objectifs en matière de sûreté.
Il est devenu manifeste que la mise en oeuvre de ces objectifs de sûreté
n'était pas une tâche aisée. Le présente communication décrit les directives
récentes de la Commission à l'intention de son personnel concernant la mise en
oeuvre de la Déclaration d'orientations sur les objectifs en matière de sûreté.

BACKGROUND

In the early days of the commercial nuclear power industry in the United States (1950's through 1970's), plant designs developed very rapidly and with them the industry's operational experience. With the rapid development of experience in operational characteristics and safety reviews, safety standards were refined with a general trend toward additional requirements. Following the accident at Three Mile Island, new requirements were imposed at a record pace. The industry, attempting to comply with an avalanche of new regulatory requirements, demanded to know how far the NRC expected to go as it developed and modified new regulations. At the same time, the public demanded to know what level of protection it could expect from reactor operations. The Nuclear Regulatory Commission's credibility with the general public was at an all time low, while the regulatory instability was at an all time high. Some of these concerns and corrective recommendations were identified in the President's Commission on the Accident at Three Mile Island.

In responding to the recommendations of the President's Commission on the Accident at Three Mile Island, the Commission stated that it was "prepared to move forward with an explicit policy statement on safety philosophy and the role of safety-cost tradeoffs in the NRC safety decisions."

The Commission initiated a project to explicitly state the level of protection which it believed appropriate to ensure public safety with regard to operation of nuclear power plants. A proposed policy statement on safety goals describing limits on risk from nuclear power plant operation was developed.

After a two year trial evaluation period, the NRC staff reported that insights from the use of probabilistic risk assessments (PRA), together with Safety Goals as a measuring yardstick, could serve to strengthen the traditional methods of arriving at regulatory decisions for a wide range of regulatory issues. With the idea that these goals might successfully be used to judge the need to develop new regulations or to modify existing ones, the Commission issued its Safety Goals Policy Statement in August of 1986, and authorized the staff to use the Safety Goals in the regulatory process.

PUBLIC PERCEPTION

The Commission believed and continues to believe that the safety goals provide a means for testing the adequacy of and need for current and proposed regulatory requirements. The safety goals express the Commission's views on the acceptable level of risk to public health from the operation of nuclear power plants and serve as an input for assessing the adequacy for future regulatory revisions.

The NRC acknowledges that there is risk in the operation of nuclear power plants, just as there is risk in all technologies and in every personal activity in which people engage. In discussing these risks, a significant fraction of the general public is confused if not alarmed at NRC's focus on the probability as well as the consequences of accidents. The general public is not accustomed to hearing discussions of risks and the likelihood of an accident occurring. Conveying to the general public the concept of relative risk and the use of probabilistic risk assessment (PRA) as a design and decision making tool is an area where the educational institutions, the industry, and the regulators have been unsuccessful.

A significant number of members of the general public expect absolute safety from nuclear power plant operation. They have difficulty comprehending the concept of probability and risk, and that absolute safety or "zero risk" is not attainable in any endeavor, nor legally required before a power plant is licensed for operation. The Atomic Energy Act refers to "adequate" rather than "absolute" protection of the public health and safety.

Since the accident at Three Mile Island and Chernobyl, the general public's confidence in the safety of nuclear technology and in the regulatory process has significantly decreased. The safety goals, along with public education and confidence in the regulatory process, is a vehicle for placing in perspective the public's perception of nuclear safety, and how the risks from nuclear power compare to other risks that an average individual faces in his or her daily activity.

DEFINITION OF SAFETY GOALS

Although the safety goals are quite simple and straight forward, they are difficult to implement.

The qualitative safety goals are as follows:

(1)   Individual members of the public should be provided a level of protection from the consequences of nuclear power plant operation such that individuals bear no significant additional risk to life and health; and

(2)   Societal risks to life and health from nuclear power plant operation should be comparable to or less than the risks of generating electricity by viable competing technologies and should not be a significant addition to other societal risks.

In order to define the phrase "risk to life and health," the Commission approved the following quantitative objectives:

-     The risk to an average individual in the vicinity of a nuclear power plant of prompt fatalities that might result from reactor accidents should not exceed one-tenth of one percent (0.1 percent) of the sum of prompt fatality risks resulting from other accidents to which members of the U.S. population are generally exposed.

(The average individual in the vicinity of the plant is defined as the average individual biologically (in terms of age and other risk factors) and locationally who resides within a mile from the plant site boundary. This means that the average individual is found by accumulating the individual risks and dividing by the number of individuals residing in the vicinity of the plant.)

- The risk to the population in the area near a nuclear power plant of cancer fatalities that might result from nuclear power plant operation should not exceed one-tenth of one percent (0.1 percent) of the sum of cancer fatality risks resulting from all other causes.

(The population considered near a nuclear power plant is taken as the population within 10 miles of the plant site.)

In establishing these goals and objectives, the Commission attempted to describe the goals in a way which would be meaningful to the public and hopefully restore its confidence in the regulatory process and in the safety of operating plants.

Underlying the safety goals objectives is the premise that the current regulatory practice of requiring compliance with the regulations ensures the basic statutory standard of adequate protection. The Commission believes, however, that the current practices could be improved to provide a better means for testing the adequacy and necessity of current requirements, as well as testing the possible need for additional requirements. The Safety Goals policy is seen as a vehicle for achieving those objectives.

IMPLEMENTATION OF THE SAFETY GOALS

In March of 1989, the NRC staff recommended to the Commission a general framework and plans for implementing the safety goals. The staff's plans for implementing the safety goals and objectives are directed toward bringing more coherence to the regulatory process, and are intimately associated with ongoing regulatory activities in the area of probabilistic risk assessment, the severe accident research activities, and implementation of the Commission's Backfit Rule.

On the fifteenth of June this year, the Commission instructed the staff on how it wanted the Safety Goals implemented. The Commission provided guidance concerning:

- Possible "partitioning" of these goals to reduce the complexity of implementation; and

- The use of the goals in making regulatory decisions more coherent and in consonance with the Commission's Backfit Rule.

SAFETY GOALS AND PARTITIONING

Fundamental to the application of quantitative safety goals to regulatory decision making is the use of probabilistic risk assessment and severe accident analytical methodologies.

PRAs, as a tool, provide only one input to the determination of whether nuclear power plant operation in the United States presents risks that are acceptable in view of the Commission's Safety Goals Policy. The Commission agreed with the advice from its Advisory Committee on Reactor Safeguards that it would not use the safety goals and PRAs to make individual plant licensing decisions. Instead, it agreed that the safety goals are to be used to help judge the effectiveness of NRC's regulations in producing plants which, as an ensemble, meet the qualitative goals as defined by the quantitative health objectives. The Commission took this approach in recognition of the limitations and uncertainties of the numerical results of PRAs. Some issues, such as, human performance and sabotage, do not readily lend themselves fully to quantitative comparisons. On a plant specific basis the Commission will continue to insist on multi-barrier, defense-in-depth designs which provide a balanced engineered approach to accident prevention and mitigation, as appropriate to the specific class of design.

With regard to a strategy for the application of PRAs and the safety goals in reaching regulatory decisions, the Commission stated:

> Probabilistic risk assessment (PRA) is used as a tool to provide measures of plant performance and overall risk to the public. Insights can be drawn from this information to evaluate the consistency of regulations with the safety goals, and to identify possible changes in the regulations that make them more consistent with the safety goals. The result of the several PRA level calculations (core damage probability, source terms, consequence estimates) as well as the results of the various internal steps within each level can be compared with certain specific regulatory requirements.

The Commission recognized that in order to relate the safety goals' quantitative health objectives in numerical terms, certain important intermediate steps would have to be identified and quantified, such as core damage frequency and containment or confinement performance. However, the Commission concluded that these need not be included in the policy statement itself.

In order to develop less complex alternative criteria which could serve as surrogates to the quantitative health objectives, the Commission approved a general plant performance guideline. That guideline states that the overall mean frequency of a large release of radioactive materials to the environment from a reactor accident should be less than 1 in 1,000,000 per year of reactor operation. This overall mean value frequency is within an order of magnitude more conservative than the Commission's

83

health objectives but provides a simple goal which has generally been accepted and can easily be understood by the public.

However, the Commission was unable to arrive at a simple definition for "large release" which would be site independent and thus eliminate the need for a Level-3 PRA. The Commission, therefore, requested the staff to develop the definition of a large release with supporting rationale which would not result in a de facto, more restrictive criterion than described by the health objectives.

The Commission also concurred with recommendations submitted by its Advisory Committee on Reactor Safeguards in five other areas by providing the following guidance to the staff:

o   As the implementation of the safety goals may require development and use of "partitioned objectives," those objectives should not introduce additional conservatism.

o   The quantitative PRA objectives should not be used as licensing standards or requirements because of analytical uncertainties.

o   The safety goals objectives should be applied to all designs, independent of the size of containment or character of a particular design approach to mitigate releases. Accordingly, for the purpose of implementation, the staff may establish subsidiary quantitative core damage frequency and containment performance objectives through "partitioning" the large release guideline. These subsidiary objectives should anchor, or provide guidance on "minimum" acceptance criteria for prevention (core damage frequency) and mitigation (containment or confinement performance) and thus assure an appropriate multi-barrier defense-in-depth balance in design. Such subsidiary objectives should be consistent with the large release guideline, and not introduce additional conservatism so as to create a de facto, new large release guideline.

o   A core damage probability of less than 1 in 10,000 per year of reactor operation appears to be a very useful subsidiary for making judgments about that portion of our regulations which are directed toward accident prevention.

o   In developing containment performance guidelines the staff should:

       -    Ensure that the conditional containment failure probability (CCFP) objective is not so conservative as to constitute a de facto new "large release guideline."

       -    Ensure that the establishment of a CCFP is approached in such a manner that additional emphasis on prevention is not discouraged. In this regard, staff should develop appropriate guidance for establishing CCFPs to address this concern and provide a uniform methodology
       - for implementing such an approach.

-   Acknowledge that it is entirely possible that a
    deterministically-established containment performance
    objective could achieve the same overall objective as a
    CCFP. Staff should be prepared to review the merits of
    such an approach (if proposed) and, if workable, accept
    such an approach as an alternative to CCFP.

The Commission approved a CCFP objective of 1 in 10 or equivalent
level of protection based on deterministic considerations for the
evolutionary designs. It further directed that:

Within a particular design class (e.g., light water
reactors, liquid metal reactors, high temperature reactors,
and heavy water reactors) the same subsidiary objective
should apply to both current as well as future designs. A
specific subsidiary objective might differ from one design
class to another design class to account for different
mitigating concepts (confinement instead of containment.)
However, the large release guideline relates to all current
as well as future designs.

The Commission emphasized that these partitioned objectives are
not to be imposed as requirements themselves, but may be useful
as a basis for regulatory guidance.

USE OF THE SAFETY GOALS FOR FUTURE REACTOR DESIGNS

In recent months, the Commission has been considering how it will
apply its quantitative safety goals objectives to future reactor
designs. It is actively reviewing two evolutionary light water
reactor designs, the ABB-Combustion Engineering's System 80+ and
General Electric's Advanced Boiling Water Reactor (ABWR). In
addition, the Commission is aware of several initiatives of the
U.S. Department of Energy to develop advanced reactors for design
certification.

The Commission is also aware of statements made by DOE's
Assistant Secretary for Nuclear Energy, William Young, that he
believes that the U.S. will both need and have some new nuclear
plant orders by the mid-1990s. Both DOE and the industry,
through the Electric Power Research Institute, are actively
working toward having the evolutionary as well as the simplified
mid-sized (600 MWe) advanced light water reactors employing
passive safety features and modular construction, certified by
1995, with the expectation that the first new nuclear plants will
be operational by the year 2000.

In the area of evolutionary and advanced reactor technology, the
Department of Energy is supporting the development and/or
certification of the following future reactor designs:

o   General Electric's Evolutionary Advanced Boiling Water
     Reactor (ABWR);

o   ABB Combustion Engineering's Evolutionary Reactor
     (System 80+);

85

o   Westinghouse's Passive Light Water Reactor (AP-600);

o   General Electric's Simplified Boiling Water Reactor
        (SBWR);

o   General Electric's Advanced Liquid Metal Reactor
        (ALMR); and

o   General Atomic's Modular High Temperature Gas-Cooled
        Reactor (MHTGR).

In addition, ABB-Combustion Engineering is developing the PIUS
and SIR Advanced Light Water Reactors, and the Canadians are
developing the CANDU-3 Heavy Water Reactor for certification in
the United States.

These advanced reactors incorporate many changes and improvements
from past designs and are presenting unique technical challenges
to the staff and Commission.

For this reason, the Commission considered separately how it will
apply the safety goals to regulatory decisions in certifying
advanced reactor designs under its new standardization
regulation.

With regard to advanced reactor design reviews, the Commission
stated:

> It is important to note that the Commission has made it
> clear in the Advanced Plant and Severe Accident policy
> statements that it expects that advanced designs will
> reflect the benefit of significant research and development
> work and experience gained in operating the many power and
> development reactors, and that vendors will achieve a higher
> standard of severe accident safety performance than their
> prior designs. The industry's goal of designing future
> reactors to a core damage probability of less that 1 in
> 100,000 per year of reactor operation (EPRI for ALWRs, GE
> for its ABWR, and ABB Combustion Engineering for its
> System-80+) is evidence of industry's commitment to NRC's
> Severe Accident policy. The Commission applauds such a
> commitment. However, the NRC will not use industry's design
> objectives as the basis to establish new requirements.

The Commission further stated that with regard to advanced
reactor licensing decisions, and in particular, when applying the
criteria provided in its regulation for a combined construction
and operating license, as well as for a certification of a design
prior to a license application, the staff should not be
constrained from proposing to the Commission new requirements
where benefits cannot be quantified in terms of risk.

USE OF THE SAFETY GOALS IN REVIEWING NEW AND EXISTING REQUIREMENTS

With regard to improving coherence in the regulatory process, as recommended by the Advisory Committee on Reactor Safeguards, the Commission directed its staff to consider the safety goals routinely in developing and reviewing regulations and regulatory practices.

To achieve this objective, the staff is developing a formal mechanism, including documentation, for ensuring that future regulatory initiatives are evaluated for conformity with the safety goals quantitative objectives.

Recognizing that the state of knowledge is such that the degree to which regulatory issues can be related to the safety goals will vary considerably, the staff's consideration of the safety goals could range anywhere from quantitative risk comparisons, involving the safety goals themselves, to a deterministic judgment that, in light of the safety goals objectives and available knowledge, or lack thereof, a given issue does or does not warrant a change to the regulations or regulatory practices.

The Commission reemphasized that the safety goals should be used when considering the need for new regulations, and that in developing and applying new requirements to existing plants, the Backfit Rule should apply.

The Commission specified the role that the safety goals should play in making backfit decisions. The Commission reemphasized that:

> There is a level of safety that is referred to as "adequate protection." This is the level that must be assured without regard to cost and, thus, without invoking the procedures required by the Backfit Rule. Beyond adequate protection, if the NRC decides to consider enhancements to safety, cost must be considered, and the cost-benefit analysis required by the Backfit Rule must be performed. The safety goals, on the other hand, are silent on the issue of cost but do provide a definition of "how safe is safe enough" that should be seen as guidance on how far to go when proposing safety enhancements, including those to be considered under the Backfit Rule.

NEW INITIATIVE

In the past, the NRC has focused its attention at operating modes which posed the most significant risk to public health and safety. Those risks were determined to occur during power operation. Through confirmatory research and regulatory requirements, the Commission reduced the potential for a significant accident during power operation to a point where "operating" modes of zero power through refueling have risen in relative significance. Consequently, the NRC recently initiated a pilot program to assess the potential for a severe accident at other than power operation and compare the calculated risk with the safety goals. Preliminary findings from this program are scheduled in approximately 2 years.

CONCLUSIONS

The safety goals express the Commission's views on the acceptable level of risks to public health and safety from the operation of nuclear power plants. The goals are also used, along with other considerations, to assess the adequacy of present regulatory requirements and need for future revisions. In establishing the safety goals objectives, the Commission attempted to describe the goals in a manner that would be meaningful to the public and help restore its confidence in the safety of operating plants.

On the fifteenth of June of this year, the Commission instructed the staff on how it wanted the safety goals objectives implemented.

Some of the major decisions included:

o   Acceptance of 1 in 1,000,000 per year of reactor operation as an overall mean frequency of a large release of radioactive materials to the environment from a reactor accident.

o   Guidance for developing "Partitioned" objectives and the requirement that they should not introduce additional conservatism such that they constitute a de facto new "Large Release Guideline," and that they not be imposed as requirements but may be useful as a basis for regulatory guidance.

o   The NRC will not use industry's design objectives as the basis to establish new requirements.

o   Guidance for applying the safety goals to future plant design reviews.

Both the industry and the regulators are faced with the challenge of regaining public confidence in the safety of reactor technology. Industry now faces the challenge of educating the general public on the technical concept of risk and restoring trust and confidence in its ability to operate nuclear power plants safely.

The Commission will continue to ensure that licensees adhere to appropriate regulations and license conditions. The Commission is also striving to improve coherence in the regulatory process. The safety goals provide the foundation from which the public and licensees more fully appreciate the basis for NRC's regulatory decisions. Implementation of those safety goals objectives in a manner which restore public confidence in both the regulatory process and nuclear safety will not be easy and will require continuing attention.

EUROPEAN SAFETY APPROACHES

E.A. Ryder
HM Nuclear Installations Inspectorate, United Kingdom

M. Lavérie
Service Central de Sûreté des Installations Nucléaires, France

K. Gast
Bundesministerium für Umwelt, Naturschutz und Reaktorsicherheit (BMU)
Federal Republic of Germany

L. Högberg
Statens Kärnkraftinspektion, Sweden

ABSTRACT

The paper reviews the European development and approaches towards safety. It concludes that comparable levels of safety are achieved and that there are similarities in approaches, although the individual country's expression of probabilistic and deterministic safety criteria and standards may vary. Current work being undertaken by European countries under the auspices of OECD-NEA, and the CEC in the fields of radiation protection and nuclear safety are reviewed.

DEMARCHES EUROPEENNES A L'EGARD DE LA SURETE

RESUME

La présente communication fait le point de l'évolution de la sûreté en Europe et des méthodes adoptées à cet égard. La conclusion qui s'en dégage est que des niveaux comparables de sûreté sont atteints et qu'il existe des similitudes dans les démarches suivies, encore que la formulation des critères et normes probabilistes et déterministes de sûreté puisse varier selon les pays. Les travaux actuellement menés par les pays européens sous l'égide de l'AEN-OCDE et de la CCE dans les domaines de la radioprotection et de la sûreté nucléaire sont passés en revue.

INTRODUCTION

The Key Issues

1.   The key issues of the European approaches to nuclear health
and safety lie in the achievement of common understanding of
safety standards, assurance and risk management.   To-date each
has evolved within the individual country's regulatory framework
and in accordance with its national practices and culture, but
it would not be correct to say that they have evolved in
isolation.   Strong bilateral and multi-lateral cooperation
patterns have evolved between a number of European Countries.
The international organisations concerned with nuclear health
and safety, notably IAEA and OECD-NEA and to a more limited
extent the CEC, each play important roles in promoting common
safety standards and providing fora for the sharing of knowledge
and experience.   If common European safety approaches can be
defined, therefore, the expression of the approaches would be
likely to find acceptance in most countries with nuclear power
programmes.

2.   The approaches to the key issues in Europe do, however show
increasing convergences of views.   Regulatory approaches;
evaluation of operating experience; systematic plant analyses
and safety reviews of older plants; improved quality assurance
in maintenance and plant modifications; the use of probabilistic
safety analyses to support assessments; the approach to human
factors by management and organisations, optimisation of the
man-machine interface, severe accident management; these are
examples of the areas where European regulators, operators and
designers are finding their philosophies, policies and practices
are converging.

Similarly, European approaches in the field of radiation
protection are internationally based on ICRP advice, and within
the countries of European Community Basic Safety Standards are
enforced under the umbrella of a European Commission Directive.

THE APPROACHES TO NUCLEAR HEALTH AND SAFETY

3.   The main approaches to nuclear health and safety in Europe
can be summarised as:-

   i)   the reduction of exposures in normal operation; and

   ii)   the prevention of accidents that could affect the
         public.

   iii) the limitation and mitigation of accident
         consequences.

Dose Reduction

4.   Common practices are found in the approach to the As Low As
Reasonably Achievable (ALARA) concept, and in general they could
be considered as state of the art. A common trend is noted of
reducing dose levels, those achieved in practice probably
significantly below European Basic Safety Standards for most
workers, and certainly for members of the public.

5.   The reduction of exposures to operating personnel has been
notable on newer plants because of the increased attention paid
to aspects of design in materials specifications and maintenance
and re-fuelling procedures. Difficulties remain on some older
plants, but the continued application of ALARA, leading to
improved procedures, development of remote handling equipment
and backfitting to meet modern standards has meant the trend in
dose reduction continues.

Accident Management

6.   The prevention of accidents that could affect the public
must be the aim of operators and regulators worldwide, but has
perhaps a special significance in Europe because of its
relatively dense population and the numbers of countries that
could be affected by a major release. Since the remote
possibility of severe accidents cannot be completely ruled out,
additional measures for risk management are taken into
consideration in European countries. Accident management
procedures which make optimal use of existing systems and
additional equipment are being developed to control even beyond
design basis plant states, or at least limit releases so that
early fatalities and long term ground contamination are
prevented.

Safety Standards and Criteria

7.   There are no formal European standards in nuclear safety
against which the plants are judged. Development of nuclear
power programmes in European countries took place at widely
different times, with different reactor types and under
regulatory and licensing systems unique to each country's
national legal framework. The expression and terminology of
safety standards criteria and requirements therefore, not
surprisingly, varies widely between countries. For example, some
countries such as the United Kingdom and Sweden, rely more on
reviews of safety cases presented by the owner/operator than
issuing detailed rules and regulations. In the treatment of
design basis accidents, and more prominently since Chernobyl in
the area of severe accidents the differences and similarities of
European approaches are best illustrated by examples from the
four countries who have contributed to this paper.

Quantitative Safety Objectives

8.   The examples chosen are those concerned with quantitative, or probabilistic, safety objectives, but are not exhaustive in terms of a full description of each. Full understanding of their application is beyond the scope of this paper, but they are illustrative of the differences and similarities of approaches used. They are nonetheless regarded as significant expressions of the levels of safety that should be achieved and in the European context have political and public importance. How safe is my neighbours reactor, and how does he prove it are questions that are asked with increasing frequency.

France

9.   The general probabilistic objective is that the design basis for a PWR should be such that the overall probability for unacceptable radiological consequences is not higher than 10^{-6} per year. For the general approach it is assumed that different classes of events (internal and external hazards, material or man- made hazards) contribute to the overall risk. If the probability of a class of adverse events capable of resulting in unacceptable radiological consequences is higher than 10^{-7} per year, then such class of events is taken into account in the design. Note that the term "unacceptable consequences" is not defined in regulatory terms but implies the obligation of taking steps to provide protection for people outside the site. Probability figures are objectives, not mandatory requirements for demonstration.

10.   Beyond the customary procedures, special procedures have been set up: "H" procedures aimed at reinforcing the preventive means and "U" procedures which are to manage an accident situation by minimising its consequences.

Germany

11.   The Safety Criteria for NPP in the Federal Republic of Germany require, that - supplementary to the traditional deterministic approach - the reliability of systems and components has to be assessed by probabilistic methods to demonstrate adequate reliability of the safety systems and a well balanced safety concept. No formal quantitative safety objectives have been established, but quantitative results of reliability analyses are used as yardsticks in current licensing and supervision. Due to the increasing maturity of methods and availability of data a growing number of reference sequences have been analysed. Current licensing analyses are equivalent to level one PSAs. Using the conservative success criteria according to the licensing requirements, the frequency of uncontrolled plant states has been assessed to about 10^{-6} per year for single event classes, summing up to a cumulative value of about 10^{-5} per year.

12. Striving for further improved protection levels, eg. by optimising preventive systems and implementation of accident management, more complete and technically detailed PSA models and plant specific analyses are considered as valuable instruments to identify means for further improvement of safety. The federal regulatory authority has, therefore, required periodic safety reassessments for all plants every ten years including an extended level 1 PSA.

Sweden

13. Sweden does not have regulatory requirements for quantitative safety objectives, but does require PSA's using plant specific information at intervals of about 10 years on each of its plants, contained within the As-operated Safety Analysis Report (ASAR). Experience so far has shown that, during the process of carrying out the ASAR programme, the Swedish utilities will implement changes in plant hardware and procedures to ensure that the calculated core damage frequencies fall below 10^{-5} per reactor operating year. In reviewing the ASAR's the Swedish regulatory authority, SKI, has endorsed this figure as a reasonable safety objective. For severe accidents, maximum activity releases are defined for mitigation measures. The objectives are that:-

-   the same basic requirements regarding the maximum quantity of released radioactive substances shall apply to all reactors irrespective of site and power,

-   land contamination which impedes the use of large areas for long periods should be prevented,

-   deaths from acute radiation disease shall not occur.

14. The requirements are considered as fulfilled if the release of fission products is less than 0.1% of Cs 134 and Cs 137 in a core of 1800 MWTh, provided that other nuclides of significance to land contamination are retained in the same proportion.

United Kingdom

15. Although not a licensing requirement in the legal sense, a licence applicant should demonstrate that his design meets the regulatory assessment levels which are expressed as:-

-   for any single accident which could give rise to a large uncontrolled release resulting from some or all of the protection systems and barriers being breached or failed, then the overall design should ensure that the accident frequency is less than 10^{-7} per reactor year,

-   the total frequency of all accidents leading to uncontrolled releases should be less than 10^{-6} per reactor year.

16. The United Kingdom has also published a consultative document "The Tolerability of Risk from Nuclear Power Stations", in which one of the proposals made implied that the chance of a severe accident giving rise to an uncontrolled release of a size capable of giving a dose of 100 mSv at 3km, and pessimistically causing eventual deaths from cancer of about 100 people should be of the order of 1 in 1 million per annum. Public reaction to the proposals was not extensive, but they were examined at an extensive Public Inquiry (Hinkley Point 'C') and the Inspector's report is awaited.

European Views on Probabilistic Safety Assessments

17. Of interest for comparison of European safety approaches is the work carried out by a CEC Working Group Task Force, reported in 1988, on the status of some European countries application of various safety goals and criteria. The study showed that European approaches to the use of Probabilistic Safety Assessment and Probabilistic Safety Criteria are similar. Probabilistic Safety Assessments, with the exception of one country, are not formal licensing requirements by the regulatory authorities. The approach is generally that PSA is uniquely valuable for the assessment of levels of risk (and hence their acceptability) and that it provides a powerful analytical tool for deeper insights into plant behaviour and reliability. PSA is further generally viewed as necessary for the production of a balanced design, meaning that dominant contributors to overall risk can be identified, and appropriate engineering measures taken to reduce them so as to avoid excessive reliance on one or two specific features. For older plants, PSA's are now generally accepted by both operators and regulators as a necessary part of periodic safety reviews, and are of particular value in decision making related to backfitting to meet modern standards. Notwithstanding the strengths of the probabilistic approach, there remains general recognition in Europe of its weaknesses in the areas of external hazards and human error for plant based assessment and the lack of accepted methodologies for computation of individual and societal risk. Regulatory and licensing decisions remain, therefore, formally based on deterministic principles.

Application of Deterministic Safety Principles

18. Similarities, differences and complexities can be detected in European approaches to probabilistic criteria, but the situation regarding deterministic criteria is even more complex. As with probabilistic criteria, deterministic criteria vary in origin in each country, either as legal requirements, sometimes imposed by regulatory authorities or self imposed by the utilities. All countries recognise and implement the key safety principles, such as redundancy, diversity, common cause failure, single failure criterion, segregation etc., but interpretation of each can differ and lead to differing design solutions.

19.   Considering the single failure criterion as an example, in all countries the installation can generally withstand in effect two independent failures in a row.   That is the result of the application of the principle, even though the requirements are expressed in differing ways.

20.   When variations in the expression of such principles are taken into account across the whole spectrum of deterministic principles, and include approaches to internal and external hazards, fire protection, human actions, maintenance and testing, leak before break arguments etc then it can be seen that starting from the same high level safety principles national requirements may lead to different design solutions.

21.   Common engineering design solutions become a question of importance in the coming decade for Europe as the countries become linked more closely under the single European market in 1993.   Plant designers, manufacturers and operators would undoubtedly see commercial advantages in being able to offer or purchase a single, licensable product in this market.   The question is therefore, how, or even should, European countries harmonize safety requirements to bring about this situation and how should regulators react?   The internationally accepted principle that each country is responsible for the safety of its nuclear installations must remain, with the corollary that through its national regulatory institutions it must set and enforce the required safety standards.   But does this preclude harmonisation of those standards without compromising on safety? If compromise is to be reached it should be in the direction of the highest agreed standards.   In this scenario regulators, designers, manufacturers and utilities all have a part to play.

22.   Each would need to gain a deeper understanding of other country's design philosophies and the application and interpretation of their safety standards, both probabilistic and deterministic, into actual design.   The examples given in this paper show that differences exist between countries and they may or may not be significant.   Certainly, actual levels of achieved safety appear to be consistent across Europe.   If there is to be a consolidated European approach, however, possibly a European design capable of being licensed in each country, then the underlying reasons for the differences, perceived or actual, need to be understood and if necessary, agreement reached to translate them into commonly accepted design solutions.

Current European Approaches

23.   European activities in the field of nuclear health and safety continue at a high level.   In the field of radiation protection the dominant issue is reaction to the revised ICRP proposals and the consequent amendment of European Basic Safety Standards.   The Euratom Article 31 expert group is charged with advising the Commission of the European Communities on any such revisions, which are likely to continue the trend for even lower

individual doses in practice. Other initiatives within the Community include a Directive to Member States on information that must be supplied to the public in the event of a radiological incident, work on radioactive contamination of animal foodstuffs, a Community wide system for early warning of radioactive releases, radiological control of outside (ie contract or itinerant) workers and radon in dwellings.

24. In the field of nuclear safety, progress is being made on Fast Reactor safety criteria, in the framework of a CEC working group which also advises on research and development work. Other research work on light water reactors is carried out by the CEC Joint Research Centre which, amongst other work is currently focussing on containment performance in severe accident conditions. Similarly in the field of light water reactor safety, a CEC Working Group which includes Sweden and Finland has carried out studies on EC Member States approaches to PSA and PSC, conducted Bench Mark exercises, studied plant safety re-evaluation, severe accident management and approaches to plant modifications. The plant safety re-evaluation study, recently published, will be of interest both to OECD-NEA in its survey on the same topic and to IAEA in its work in producing a safety guide. Each state compared its approaches against three criteria - does the plant still meet its original design safety requirements; does it have any life limiting features and how does it meet modern standards? The responses showed many similarities of approach bearing in mind the different regulatory regimes and types and ages of reactors in Europe.

25. In other areas of nuclear safety the work of OECD-NEA, could by virtue of the many European member states participating be viewed as encompassing European safety concerns. These include severe accident phenomena in reactor containments, containment system performance, plant ageing, the use of computers in plant safety management, operational data patterns and trends, uses of PSA, human factors studies, accident management, periodic safety reviews and many other technical issues.

CONCLUSIONS

26. There is no single European approach to nuclear health and safety; rather it could be said that there are as many approaches as there are countries. But this does not mean that the differences in approach are necessarily significant when overall levels of achieved health and safety are considered. Given the widely different natures of European countries power programmes, types and ages of plants, legal systems and national practices many similarities of approach emerge on deeper study. This paper, by use of brief examples, has illustrated that the philosophies, policies and standards are similar, but that there may be differences in interpretation and usage.

27. The European challenge in the coming years is to meet public expectations in nuclear health and safety, to be able to demonstrate convincingly that each country's nuclear safety standards are comparable one with another and are of the highest order, and importantly to show that these standards are translated into safe and reliable operation.

28. With comparatively few suppliers of nuclear plant, the question of uniformity, or harmony, of standards and design has significant economic implications, and regulators, while, of course, giving priority to safety aspects, cannot be immune to this debate. So far, harmonisation of standards has evolved through exchanges, both bilateral and multi-lateral, and through the auspices of international organisations tasked to promote nuclear safety. The exchanges lead to better understandings, but clearly more is needed. Europe is not yet at the stage where a single design is known to be licensable in all countries without modifications.

THE PRINCIPAL NUCLEAR SAFETY POLICIES IN JAPAN

Hideo Uchida
Chairman, Nuclear Safety Commission
Kasumigaseki 2-2-1, Chiyoda-ku, Tokyo 100, Japan

ABSTRACT

The basic policy for nuclear safety in Japan is to give first priority to the prevention of occurrence and propagation of abnormal events rather than to the mitigation of the consequences of incidents and accidents. Preventive maintenance is an important countermeasure for aging degradation. The seismic design policies and other specific design concepts relevant to safety are discussed. It is recognized that PSA is important and indispensable for comparing safety problems. However, the applicable limitations associated with PSA should be addressed. The interface issues in Health Risk and Nuclear Safety should be discussed by concerned experts.

GRANDS AXES DE LA POLITIQUE DU JAPON EN MATIERE DE SURETE NUCLEAIRE

RESUME

Au Japon, la politique fondamentale en matière de sûreté nucléaire doit privilégier avant tout la prévention de la survenue et de la propagation des événements anormaux plutôt que l'atténuation des conséquences des incidents et accidents. La maintenance préventive constitue une importante contremesure face à la dégradation due au vieillissement. L'auteur examine les politiques suivies en ce qui concerne la conception parasismique et d'autres notions spécifiques de conception ayant trait à la sûreté. Il est reconnu que l'EPS est importante et indispensable lorsqu'il s'agit de comparer les problèmes de sûreté. Il conviendrait, toutefois, de traiter les limitations actuelles liées à l'EPS. Les experts concernés devraient se pencher sur les questions d'interface entre les risques pour la santé et la sûreté nucléaire.

1. Present status of nuclear power development in Japan

At present in Japan, thirty-eight commercial nuclear power plants are in operation. Their total generating capacity is 30.38 GWe which represents approximately 18.4% of the total generating capacity in Japan. In addition to these operating plants, 12 plants with a total capacity of 12.0 GWe are under construction. Three plants, with a total capacity of 3.54 GWe, including two ABWRs (Advanced BWRs), are under review by licensing authorities. Therefore, a total of 53 units, 46.6 GWe, will be in operation by the middle of the 1990's. A Long-Term Program for Development and Utilization of Nuclear Energy, published in June 1987 by the Atomic Energy Commission, estimates nuclear power capacity in Japan will be at least 53 GWe (approximately 25% of total capacity) by the year 2000. The majority of the Japanese people recognize that the present nuclear power plants are highly safe and reliable because of the excellent performance of these plants. However, considering the increasing antinuclear movement in Japan which have been strengthened by the effects of the TMI-2 and Chernobyl accidents and environmental concerns in general, it will not be easy to achieve the development target of the Atomic Energy Commission. All operating commercial reactors in Japan except for one gas cooled reactor are LWRs which were developed in the U.S.A. Two reactors have been developed in Japan. Fugen (165 MWe), a prototype heavy-water-moderated, light-water boiling reactor (pressure-tube type), has been in operation since 1979. Monju (280 MWe), a prototype liquid-metal-cooled fast breeder reactor, is under construction and is aiming at initial criticality in 1992.

The development and use of atomic energy in Japan have been pursued in accordance with the basic policy described in the Atomic Energy Act, that is, "nuclear energy is only for peaceful purposes, and safety assurance is the first consideration." Japan has had no accidents which involved meaningful release of radioactive materials and which had a detrimental effect on the public. The average availability factor for nuclear power plants has been higher than 70%. The safety and reliability performance achieved by Japan gives Japan an excellent reputation within the international nuclear community.

Discussions about the probabilistic safety assessment (PSA) and severe accidents issues have persisted since the TMI-2 accident. It may be necessary that experts should now reach a consensus on "how safe is safe enough". The principal nuclear safety policies and concepts of concern in Japan, including the author's personal opinions, are presented herein. To continue successful development of nuclear power in Japan, it is necessary to instill and promote an accurate understanding of nuclear safety concepts in the Japanese people.

2. NPP operational experience

The safety regulations for Japanese nuclear facilities, including the issuance of an establishment permit, the issuance of an operating license, inspection during operation and reactor decommissioning are executed by a single competent authority according to the type of the facility and its status. Consistent administration of safety regulations is the foundation of this policy. The Ministry of International Trade and Industry (MITI) is the competent authority for the commercial nuclear power stations, and the

Ministry of Transport is the competent authority for commercial marine reactors. The Science and Technology Agency (STA) is the competent authority which administers research reactors, reactors under development and all nuclear fuel cycle facilities.

The Nuclear Safety Commission (NSC) was established in 1978 as an advisory organ to the Prime Minister. The commission has a position independent of the administrative organizations and, of course, the industry and is responsible for determining the fundamental policies related to nuclear safety regulation.

Japan has operated reactors for 24 years and, as to March 31, 1990, has approximately 400 reactor-years of NPP operational experience. The average plant availability factor for each of the past seven years was 70% or higher. For example, it was 77.1% for the calendar year 1987, 71.4% for 1988 and 70.0% for 1989. No reactor accidents or incidents that resulted in a release meaningful amount of radioactive materials outside a plant site, have occurred in Japan during these 24 years. The number of events that induced an unscheduled automatic shutdown (scram) during reactor operation due to a loss of off-site power, an operator error or other unusual occurrence or trouble in systems or equipment was less than 0.2 a year per reactor on the average. For example, the total number of scrams was four for fiscal year (FY) 1988 (36 reactors in operation) and one for FY 1989 (37 reactors in operation). In FY 1989, 22 events were reported in accordance with the law; one scram during normal operation due to a feed-water pump trip caused by human error during testing, 10 problems identified during normal operation which were followed by manual shutdown for inspection and repair and 11 problems diagnosed during periodic inspection which required repairs. Besides these, minor problems were reported in 13 cases.

The main problems which occurred in early Japanese plants were intergranular stress corrosion cracking (SCC) for BWRs, steam generator (SG) tube leaks for PWRs, fuel failures (leaking, bending and deformation) and the miscellaneous failures associated with the early stages of reactor development. For these problems, substantial experimental investigation has been conducted.

Problems concerning fuel have been solved by means of design improvements and material improvements identified by in-reactor fuel irradiation experiments with the Japan Materials Testing Redactor (JMTR) and reactors in foreign countries. Leaking fuel rods have seldom been observed in recent years.

The SCC problems mainly associated with stainless steel pipes in BWRs have largely been solved by means of design improvements to prevent excessive stress and to eliminate locations having stagnant flow, changing pipe materials to low carbon stainless steel, careful water treatment to reduce oxygen and chloride and improvements in welding methods by reducing thermal heat input and releasing residual stress for welded portions by applying a special in situ thermal treatment. The SCC observed at the flexible pins of control guide tubes of PWRs has been eliminated by improving the design and the manufacturing processes.

Steam generator tube problems were mostly eliminated by improvements in the chemical treatment of the coolant and improvements in techniques for pipe expansion at the tube sheets. Defects observed on SG tubes are all repaired by careful plugging or sleeving before leakage starts. As a result, leakage of primary coolant from the primary to the secondary side of SG tubes now rarely occurs. Corrosion failure of steam generator tubes in a few reactors of early construction is the only problem unresolved.

Careful filtering and chemical treatment of primary coolant and the use of low cobalt stainless steel for pressure boundaries has maintained the radioactive materials in the reactor coolant at a very low level. These measures have reduced radioactive exposure doses to nuclear plant workers and radioactive material releases to the environment to negligibly low levels as well. The radiation exposure for the public during normal operation is negligibly low, estimated to be less than one-tenth of the reference dose of 0.05 mSv/year/site which is recommended for operation and management of commercial light water reactors.

Because the SCC and SG tube leakage problems have been essentially solved, most unusual events that have occurred in recent years have been due to troubles in instrumentation/control equipment, inadequate quality assurance (QA) during construction, commissioning or maintenance, or as the result of aging degradation. Emphasis is placed on preventive maintenance where plant maintenance efforts are earnestly exercised to detect symptoms of failures and discover equipment aging phenomena in the early stages and to implement appropriate replacement or repairs.

After a commercial nuclear power plant begins operation, it is required by law that equipment, components and systems relevant to safety be inspected by MITI approximately every 13 months. Refueling normally occurs during this periodic inspection. The items which are covered by the periodic inspection are predetermined: an inspection of almost all fuel bundles, an in-sevice inspection (ISI) of the primary coolant pressure boundary, an eddy current test (ECT) of almost all steam generator tubes, etc. The electric utility conducts, of its own accord, inspection, testing, modification and replacement of equipment during the periodic inspection. The periodic inspection improves the safety and reliability of the plant because it uncovers equipment defects. Thus, failures, abnormal events and accidents are prevented.

When a newly constructed commercial nuclear power plant passes MITI's final inspection, MITI issues an operating license which authorizes the start of operation and permits operation of the plant for its life. The MITI also permits the plant to resume commercial operation only after a comprehensive operational test is successfully conducted during the final stage of periodic inspection, conforming that the plant as whole is qualified to a technical standard. This rule has the same effect as renewing the operating license every year.

The existing facilities, systems and components have been designed and manufactured to have service lives of 40 years or more. A plant life extension program and research on aging problems are being conducted in Japan. Measures to mitigate plant aging are being studied with the expectation of extending plant life to 50 or 60 years.

3. The basic concept and policy of nuclear safety assurance

The development of nuclear power plants in Japan started in 1963 when JPDR (Japan Power Demonstration Reactor) began operation. The first commercial nuclear power plant in Japan was a carbon dioxide gas cooled reactor (166 MWe) introduced from the U.K. in 1966. Since 1970, primarily light water reactors developed using technology introduced from the U.S.A. have been constructed in Japan. Therefore, the basic Japanese policy fo assurance of nuclear safety and safety standards has drawn much substance from the regulations and policies of the U.S.A. However, based on nearly thirty years of experience in manufacturing nuclear components and in building and operating nuclear power plants , policies and standards have evolved to detailed procedures and criteria which meet the requirements of the administrative system, society, technology, etc. in Japan. International trends and consensuses have always been addressed since the concerns about nuclear safety should have internationally common identifications. For example, the codes of practice and the safety guides established by the NUSS program of the IAEA and the resolutions of OECD-NEA, the ICRP and the IAEA have been largely incorporated into the Japanese safety standards and practices. The findings of investigations and examinations from the TMI-2 accident, the Chernobyl accident and other occurrences have also been integrated.

The basic safety policy for the nuclear facilities in Japan is to protect the health and safety of the public and nuclear plant workers from undue risks associated with radiation and radioactive materials which may result from normal operation, abnormal occurrences or accidents at nuclear power plants. The basis of safe design of nuclear facilities corresponds to internationally recognized principles; that is to conform to highly conservative designs based on the principles of defense-in-depth, multiple barriers, single failure criterion, etc., and which give first priority to measures which prevent abnormal events or accidents.

To implement this policy and ensure the safety of nuclear facilities, it is necessary first to design, manufacture and construct them using high-level technology. It is most important that the quality of hardware, components and equipment, be maintained at high-level standards. It is also necessary to reliably manage operation of nuclear power plants through careful quality assurance and by skilled operators. Supervisors and operators of nuclear power plants have primary responsibility for the safety of the plant. However, all people involved in the design, manufacture, construction, operation and the regulatory bodies should strive to promote the safety culture in each category.

Emphasis must be placed on the need to always pay attention to indications of leakage, wear, corrosion, vibration and other minor abnormal symptoms which may be detected during normal operation or inspections. Appropriate responses to these symptoms contributes toward avoiding abnormal events or accidents. When defects are found in equipment or systems, it is important to thoroughly examine the defect for a possible common cause failure candidate.

The Japanese basic policy for radiation protection has drawn much guidance from ICRP recommendations and these have been largely incorporated in the Japanese criteria. The current Japanese Radiation Protection Law was revised in 1989. It is based on ICRP Pub 26. The maximum permissible exposure dose for a person is 1 mSv/year for the public and 50 mSv/year for workers at nuclear facilities, as an effective dose equivalent. Referring to the Paris Statement, a member of the public may be permitted 5 mSv/year instead of 1 mSv/year in specific cases. The emergency permissible dose for workers was changed to 0.1 Sv in effective dose equivalent. Concerning the practical reference dose recommended by the ALARA concept, the guideline was revised to express that exposure to the surrounding residents through both liquid and gaseous radioactive materials released from the light water reactors in a power station shall not exceed 0.05 mSv effective dose equivalent per year.

An important policy for safety countermeasures against accidents at nuclear power plants is to ensure the safety of the public even if a serious accident occurs. This concept is that "the surrounding residents do not receive nonstochastic effects of radiation (no early fatalities) even if a highly unlikely major accident is hypothesized to occur, and the risks from the stochastic effects of radiation which the public receives are mitigated as low as reasonably achievable."

The distance of nuclear power plants to the surrounding residents is evaluated for postulated major accidents for this purpose. Such postulated major accidents are called site evaluation accidents (SEA). The SEAs are included in the category of beyond design basis accidents and should encompass all radiological consequences to the public by design basis accidents with a sufficient margin.

The site conditions in Japan for ensuring safety with respect to such a postulated major accident have been established in the Examination Guide of Reactor Siting and Guidelines for Interpretation in Their Application.

The basic concepts of the guideline are as follows:

To protect members of the public proximate to the site from radiation injuries and hazards in case of a reactor accident, two accidents are postulated to evaluate the isolation distance of the reactor from the surrounding residents. These are a serious accident which is considered to be foreseeable from a technical viewpoint, and a highly unlikely accident which is considered improbable from a technical viewpoint. The former is called a major accident, and the latter is called a hypothetical accident.

When a major accident is postulated, the area in which the predicted exposure of the public will exceed 0.25 Sv whole body and 1.50 Sv thyroid (child) is to be a nonresidential area. When a hypothetical accident is postulated, the area in which the predicted exposure of the public will exceed 0.25 Sv whole body and 3.0 Sv thyroid (adult) and which is outside the nonresidential area is to be a low population zone. In this situation the collective whole body dose of the surrounding public is to be less than 20,000 person-Sv. These doses, called reference doses, are used as a standard when evaluating the isolation distance of reactor facilities from the public. It should be noted that they are not defined as permissible exposure doses.

In principle, what is called a remote siting policy has not existed in Japan. Nuclear power plants in Japan are located on the shore of the ocean and are constructed on firm bedrock from the point of seismic design. In Japan, the nonresidential area and the low population zone necessary for siting evaluation are in most cases included within the site boundary, usually within a 400-1000 m radius from a reactor. Evaluation of the collective whole body dose of the surrounding public for a hypothetical accident is considered an acceptable method to evaluate the societal risks for a densely populated country like Japan.

Taking a LOCA as an example of a major accident, off-site power loss and the single failure criterion are applied, and the function of engineered safety features such as ECCS (Emergency Core Cooling System) are taken into account. For a hypothetical accident, the function of ECCS is neglected, and the hypothesized release into the containment of fission products is assumed to be 100% of the rare gases and 50% of the iodine of the core inventory. However, the integrity of the containment to cope with the temperature and pressure generated is presupposed. In addition to

LOCA, a rupture of a steam generator tube for a PWR and a break of the main steam line outside the containment for a BWR are evaluated as site evaluation accidents.

Based on the safety concepts mentioned above, it is clear that the safety of a nuclear power plant is ensured by design, manufacture, construction, and operation, based on high technology. In other words, safety of a nuclear power plant should be ensured entirely within its site. The following concept was determined after a full discussion at the NEA/CSNI in 1980; technical problems concerning design, manufacture and construction, and problems concerning site conditions are given consideration as problems separate from countermeasures against emergency. They are important and closely connected within the entire safety problem. However, countermeasures against emergency are not within the framework of site evaluation accidents.

The nuclear emergency procedures refer to countermeasures and actions taken as precautions by administrative organizations (the central and local government) for an emergency at nuclear facilities. These operations would play a partial role in accident management in mitigating the off-site radiological consequences and residual risks of a severe accident which leads to a significant release of radioactive material. Countermeasures for emergencies concerning nuclear accidents in Japan are based on the Fundamental Law for Disaster Management which is applied to earthquakes, typhoons, tidal waves and all other disasters. This law prescribes that the administrative organizations should be responsible for establishing emergency procedures and implementing them when necessary.

Electric utilities are responsible for establishing and implementing on-site emergency procedures. The NSC with its subordinate, the Emergency Technical Advisory Body, is responsible for establishing the fundamental policy for emergency planning and providing the government with the technical knowledge and information required to implement the plan. The NSC established the Nuclear Emergency Guideline composed of the technical information required for emergency planning. This Guideline recommends that an area within an 8-10 km radius of the nuclear power station should be designated a zone for emergency planning and for which an emergency plan specific to the nuclear power plant should be prepared, taking into consideration the natural and social conditions around the site.

Of the radioactive wastes generated in the nuclear power plant, the gaseous and liquid wastes are controlled by specific requirements and may only be released to the environment in accordance with established standards. Liquid wastes are reduced in volume by evaporation. Combustibles of solid wastes are burned in an incinerator to reduce their volume. Sludges remaining after evaporation and incombustibles of solid wastes, together with radioactive solid wastes, are packed in drums and solidified with cement, asphalt, plastic, etc. These drums are presently storaged on-site at each plant. The level of radioactivity in the solidified radioactive waste drums is low. The equivalent of about 500,000 drums, each with a capacity of 200 liters, are storaged throughout the country. The total storage capacity available is 785,000 drums. The spent control-rod absorbers and some other solid wastes are storaged separately on-site.

These drums containing low-level solidified radioactive waste are scheduled for shallow land disposal in the near future. The disposal location is Rokkasho-Mura, Aomori Prefecture. Safety studies concerning such land disposal are in progress. They are addressing artificial shielding, natural materials shielding, natural conditions, etc. The level of radioactivity is expected to decay over a long period of time (about 300 years) to a degree that radiation control at the disposal location will not be required.

4. Examples of basic design policies

While the basic policies for safe design of the LWR facilities in Japan are not greatly different from those conforming to the common international consensus, as noted previously some variances exist due to operating experience and the special characteristics of nature, the society and administration, etc. in Japan. As described earlier, the fundamentals of safe design lie in giving priority to the prevention of occurrence and propagation of abnormal events rather than to the mitigation of the consequences of these accidents. Some examples of basic design policies are presented.

(A) Large natural forces are expected to occur in the vicinity of the site of a nuclear power plant. These include earthquakes, tsunamis (high sea waves), typhoons, and floods. The basic approach to nuclear power plant design is to assume natural forces larger than the largest events experienced at the vicinity of the site in history. This concept is described in the IAEA Safety Series No. 50-C-S as "the radiological risks associated with external events should not exceed the severity of radiological risks associated with accidents of internal origin". In such design considerations, the seismic design against earthquakes must have the greatest importance in Japan.

A more rigorous approach is taken in designs when addressing seismic forces than when addressing other natural forces. The sites of nuclear power plants in Japan are selected at seismically stabilized zones which are distant from highly active faults. Geologically hazardous areas are excluded. The foundation of a nuclear power plant is constructed on firm bedrock which originated in the Tertiary geological era or earlier.

The seismic design is performed so that a reactor accident which would lead to large releases of radioactive materials could not be caused by either the largest historical earthquake or a much more severe earthquake which is extremely unlikely to occur but can be postulated from engineering observations based on seismotectonics, etc. An earthquake ground motion which could shake the reactor facilities by an earthquake postulated for this purpose is called a basic design earthquake ground motion.

Equipment, systems, etc. of a nuclear power station are classified into seismic Class A, B, and C according to the importance of seismic design. These are grouped according to the degree of possibility of causing a release of radioactive materials to the environment due to the loss of function. Class C includes items equivalent to usual industrial equipment and facilities. Class A includes the reactor coolant pressure boundaries, the facilities for spent fuel storage, the emergency reactor shutdown system, the decay heat removal system, the long-term core cooling system, the ECCS, and the containment facilities. Especially important equipment, piping and facilities of Class A are designated as Class As items. Class B includes items of equipment, piping, and facilities not included in either Class A or C.

The seismic design is analyzed dynamically or statically. For Class C, static analyses are conducted in accordance with the Building Standard

Law for general buildings, using the horizontal-shearing force coefficient (0.2 standard) specified therein. For Class A and Class B, static analyses are conducted using horizontal-shearing force coefficients 3.0 and 1.5 times that used for Class C, respectively. Class A and As are mainly analyzed dynamically. Class B facilities which may resonate with the earthquake vibrations are analyzed dynamically also. The basic design earthquake ground motion used in dynamic analyses of the seismic design of the reactor facilities is based on basic earthquake ground motion postulated at the free surface of the base stratum of the site. Two types of basic earthquake ground motions are postulated: an earthquake ground motion S1 caused by the strongest earthquake considered to have a possibility of occurring from an engineering viewpoint (called the maximum design basis earthquake) and an earthquake ground motion S2 caused by the strongest earthquake which can be assumed to take place based on the engineering observations of the seismotectonics around the site, etc. (called the extreme design basis earthquake).

The basic earthquake ground motion is defined based on the design earthquake. The magnitude and location are postulated, taking into account the seismotectonics, active faults associated with earthquakes, the historical records of earthquakes which occurred in the region surrounding the site, etc. The basic earthquake ground motion is expressed by a response spectrum or a time-history simulated seismic wave with maximum amplitude, frequency characteristics, durations and changes with time taken into consideration. The maximum amplitude of the earthquake ground motion is normally indicated by a velocity.

Historical records are surveyed for descriptions of earthquakes that occurred within about 200 km of the site. Those earthquakes of Seismic Intensity V (called a strong earthquake) or higher by the Meteorological Agency's intensity scale (Intensity 0-VII) are examined closely. Intensity V is equivalent to an Intensity of 7 3/4 on the Modified Mercalli Intensity Scale. An extreme design earthquake associated with active fault displacement in the vicinity of the site which is postulated to occur from seismotectonic considerations usually ranges to Magnitude 7.5-8.5. For an extreme design earthquake, an earthquake of at least Magnitude 6.5 is postulated to occur about 10 km directly under the reactor facility center.

From investigation and research of historical earthquake records, seismology and the seismotectonic structure of Japan, the following concept is drawn: an earthquake will occur by releasing the energy accumulated within the zone concerned. The zone will have an infinite area, and the recurrent period which is inherent to the zone will be some thousands of years or less. From this concept it is concluded that there is an upper boundary for the maximum magnitude of the potential maximum earthquake estimated to occur in the vicinity of sites in Japan. An extreme design earthquake corresponds to the potential maximum earthquake mentioned above or is larger. It seems unreasonable to assume, based on so-called probabilistic analyses, that a larger earthquake would occur with a recurrent period greater than some thousands of years.

The foundation of a reactor facility must, in principle, be constructed on firm bedrock formed in the Tertiary geological era or earlier. Each building must have a rigid structure. To meet the design requirements, Class A facilities must be designed to withstand the loads due to S1, and Class As facilities must be further able to maintain their safety functions against S2 loads when combining dead loads and normal operational loads. For example, the containment structure is designed to maintain the design leak rate, although plastic deformation rather than elastic deformation in some parts is allowable to resist the loads due to S2.

(B) Examples of the fundamentals of safe design, excluding seismic design, are as follows:

(1) Anticipated operational occurrences must be terminated before the core is damaged. Further, the core must be placed in a condition which allows return to normal operation. For events going beyond the anticipated operational occurrences during operation but within design basis events, the whole-body exposure of the surrounding public must be evaluated to be less than 5 mSv, except for events whose occurrence probabilities are very low.
(2) Negative reactivity feedback characteristics must be held for transient changes in power over the entire operating range, with the moderator temperature coefficient, the Doppler coefficient, the moderator void coefficient, the pressure coefficient, etc., all taken together. (3) The fuel enthalpy increase of a postulated reactivity insertion event must not rise over 230 cal/g UO_2. (4) The BWR containment is charged with inert gas and is also equipped with flammable gas control systems (FCS). (5) The large containment of a PWR is designed so that an FCS can be installed when necessary. An ice condenser containment, having a smaller volume, is always equipped with two FCSs. (6) The reactor must be safely shutdown, and core cooling must be ensured after shutdown, even with a station blackout up to 30 minutes. This requirement results from consideration of the operational experience of the power source reliability. The frequency of loss of transmission power for more than 30 minutes of the high voltage transmission lines above 187 kV is 1.26×10^{-4}/km/year, the probability of off-site power loss is about 1×10^{-2}/year and the failure rate for starting of the emergency diesel generators is less than 1.2×10^{-3}/demand in Japan.

As stated previously, the radiation exposure of the surrounding members of the public through gaseous and liquid radioactive materials released from a light water reactor power plant during normal operation is limited to 0.05 mSv/year per site from the standpoint of "ALARA" in Japan. The processing and disposal of radioactive materials are given careful consideration and the radiation exposure is, in fact, lower than 0.005 mSv/year at every site in Japan. The following examples are practices and accomplishments in Japan: (1) Hydrogen in radioactive gases extracted from the primary coolant of PWRs is removed by a hydrogen recombiner. (2) Radioactive gases from PWRs and BWRs are passed through a charcoal filter bed to reduce their radioactivity. For PWRs, the activity of the radioactive gases may be decreased further by holding them in a decay tank and released them to the environment after the activity is reduced sufficiently. (3) All radioactive liquid is collected, reduced in volume by evaporation, filtered and stored to decay radioactivity. (4) Clean, live steam generated by a house boiler is used as turbine shaft seal steam for BWRs. Exhaust gases extracted from the condenser are processed when the vacuum of the condenser is increasing during start-up. (5) There are almost no fuel rods in the core which have defects such as leakage or bending during operation. Steel materials with low cobalt content are utilized and the water chemical quality control of the reactor primary coolant is implemented conscientiously. These factors contribute to a low level of radioactivity in the primary coolant with a resultant reduction of occupational exposure during inspection, repairs, and so on.

5. Severe accidents

A "severe accident" is one which exceeds the design basis event or does not fall within its concepts. In a severe accident, either proper core cooling or reactivity control is not maintained by methods or function of equipment and systems which are assumed in safety design evaluations. Thus, a severe accident would result in failure of core cooling or loss of reactivity control which could lead to severe core damage and could cause considerable radiological effects on the environment. However, a severe accident is considered not to occur from an engineering viewpoint.

The seriousness of a severe accident depends on the degree of fuel damage and on the degree of loss of containment integrity. Countermeasures to be taken for the severe accident should differ according to the evaluation results in connection with radioactive materials released to the environment, the radiation effects to the general public and the potentiality/probability of its occurrence.

The design basis events are accidents or events which are postulated to confirm the adequacy of safe design from a viewpoint of safety assurance of a nuclear facility. The regulatory bodies must examine the principal design concepts of the nuclear power installations related to design basis events in advance of the issuance of the plant's establishment permit. The design basis events are postulated to be caused by a single failure of equipment or systems of the facilities or by a single operational error caused by operators. For each design basis event, conservative analyses and evaluations are completed, assuming a single failure of the system relevant to the safety function and loss of either off-site power or emergency power. Based on these results, the reactor facilities are designed and operated so that the public will not suffer severe injuries or hazards through radiation. The main responsibility of the regulatory bodies is to assure these safety objectives.

As a result of the TMI-2 accident, it is recognized that severe incidents can occur which would produce significant damage to the core, including partial core melting. However, it is also recognized that the conservative design margin and appropriate countermeasures finally cooled the degraded core, confined radioactive materials in the containment and terminated the TMI-2 accident. The ultimate safety goal in the design and operation of nuclear power plants should be to achieve the following:
(1) The reactor must be securely shutdown; (2) Even if the core is partially damaged, in-vessel cooling of the core must be established (so that the core is contained within the pressure vessel); (3) Radioactive materials must be confined within the containment so that significant releases to the environment are prevented.

The results of the TMI-2 and Chernobyl accidents have increased the importance of studying both severe accidents and the countermeasures against such accidents. These studies should give priority to the following goals: (1) Accumulating knowledge of sequences of incidents which lead to a severe accident, that is, accident scenarios; (2) Evaluating the role of the safety margin included in the current design for preventing a severe accident and mitigating its consequences; (3) Countermeasures to prevent the propagation of an accident and to mitigate its consequences, that is, countermeasures for accident management; (4) Measures to restore the functions of failed components and to minimize their effects on the accident; and (5) A probabilistic safety assessment (PSA) of severe accidents which incorporates all the above and an estimate of the potential risks. It is necessary that studies on severe accidents should be conducted in association with PSAs. The containment response to loads in a severe accident, the ultimate strength of the containment, and its safety function are important and must be prudently examined in Japan. In the course of this examination using Level 2 PSA, careful consideration should be given to the merits and demerits of employing measures, such as improved containment venting systems, to prevent containment failure caused by overpressurization and overheating.

6. Probabilistic safety assessment (PSA)

Safety regulations in Japan require, with a few exceptions, that safety evaluations be based on deterministic methods. However, PSAs are applied to supplement the deterministic evaluations. For example, probabilistic evaluations are conducted for aircraft crashes and turbine missile accidents at nuclear power plants. If the estimated probability of occurrence of an effect which may impair safety is less than 10^{-7}/year, design countermeasures are not required. It should be mentioned that aircraft flight below a specified height directly over or in its vicinity of the nuclear facilities is not allowed.

The most respectful appreciation should be expressed for the probabilistic nuclear safety studies developed in WASH-1400, the German Risk Study, etc. These PSAs are indispensable for the integral assessment of generic safety issues and the synthetic evaluation of severe accidents associated with nuclear power plants, particularly the LWRs developed in the industrialized countries of the West. However, there are so many uncertainties, assumptions and expert's judgments included in these PSAs that the numerical expressions of probability and consequence of accident should be used only as a tool for comparing problems concerned with nuclear safety.

The results obtained by PSAs associated with nuclear power plants are justificable and understandable only when the technical and analytical methods concerned are employed within the framework developed through WASH-1400 etc., and when the preconditions, expert's judgments and uncertainties which are included have international consensus of usage. For example, probability of occurrence of a highly unlikely accident is obtained, using event trees and fault trees, from the product of the unreliabilities or estimated failure rates of the systems and equipment concerned. Therefore, the probability figure for an accident does not express its actual likelihood of occurrence. Instead, it expresses the level of "unsafety". It is certain that the probabilities and consequences obtained from a PSA cannot actually be verified. The probability has a meaning which expresses the likelihood of occurrence of an accident only when its occurrence is assumed in order to perform a nuclear safety/risk evaluation. It must be understood that there are limitations for the use of the analytical results of PSA technology. The consequence of an accident expressed in a PSA is controversial as will be mentioned later.

The Japanese owners' group and other research institutes have conducted a PSA for Japanese nuclear power plants. The special committee for PSA established under the NSC has completed a preliminary evaluation of the results equivalent to a Level 1 PSA. The committee reported that the severe core damage probability for Japanese LWRs is far less than 10^{-5}/RY.

In this study "severe core damage" is defined as the state in which fuel cladding surface temperature is analytically estimated to exceed 1200°C in some part of the core. The principal contributors to core damage probability are a LOCA for a BWR-3, transient incidents such as loss of off-site power for a BWR-4 and -5, and a LOCA for PWRs. These figures are obtained using as much data as possible from Japanese power plants' operational records, such as the loss of off-site power and transients which occurred, taking into consideration the uncertainties included in the data and in the analytical methods. Based on this study, the probability targets for severe core damage accidents shown in "Basic Safety Principles" of the IAEA INSAG are considered to be satisfied in Japan. The severe core damage probabilities are about one order of magnitude lower than these of nuclear power plants outside of Japan. This results mainly from the use of actual data on reliability concerning the loss of off-site power loss, the operability of diesel generators, and transients experienced in Japan, as mentioned above. The reliability of off-site power depends on specific circumstances and problems associated with the site conditions and the electrical supply network system for each nuclear power station. Therefore, it seems reasonable to consider the loss of offsite power as an external event. Concerned Japanese experts are proceeding to the next step, Level 2 PSA, in which the probability and time sequence of a large release of radioactive materials will be studied. The integrity of the containment in the case of a severe accident will address issues related to venting systems. In these studies, as much data as possible from Japanese power plants' actual operation records will be used to provide a more rational evaluation.

If the estimated probability of initiation of an event, either internal or external, is below a certain limit, for example 10^{-7}/RY, such events should be eliminated from the overall PSA study to be used for safety regulation.

As discussed previously, the probability of severe core damage due to an external event, for example an earthquake, must be far less than that of accidents associated with internal events, such as loss of coolant due to pipe rupture. Also as mentioned before, in Japan it seems unreasonable to assume a probabilistic relationship between the magnitude of an earthquake and its recurrent period beyond some thousands of years. It is understood that there is an upper bound of the magnitude of an earthquake estimated to occur in the vicinity of a nuclear power plant in Japan. The extreme design earthquake is more severe than the upper bound earthquake. The aim of PSA for an accident initiated by an earthquake is not to evaluate the specific incidents due to an earthquake more severe than the design earthquakes. The aim is to study weak points of components and systems related to safety to cope with seismic force, and to obtain needed reliability data to study analytical methods to establish the basic earthquake ground motion and so on.

7. Safety goal

The IAEA Basic Safety Principles state "... the level of risk due to nuclear power plants does not exceed that due to competing energy sources, and is generally lower." The USNRC states in its Safety Goal Policy "The risk to an individual or to the population in the vicinity of a nuclear power plant site of prompt fatalities that might result from reactor accidents should not exceed 0.1% of the sum of prompt fatality risks resulting from other accidents...." It further states "The risk to an individual or to population in the area near a nuclear power plant site of cancer fatalities that might result from reactor accidents should not exceed 0.1% of the sum of cancer fatality risk resulting from all other causes."

The internationally accepted concept for acceptable risk due to nuclear accidents – from both prompt fatalities and latent fatalities – is that this risk should be sufficiently lower than other risks which the public faces in daily life. From these considerations, it is understandable that the safety goal for a nuclear power plant should aim at a fatality risk due to nuclear accidents lower than about $10^{-6} - 10^{-7}$ per reactor year.

The nuclear accident risk is defined as the arithmetic product of the probability of an accident or an event and the adverse effect it would produce. It is usually expressed in terms of the number of deaths that would result or the fatality rate. The adverse effects associated with nuclear accidents can be expressed as follows:

(1) The radioactive materials released to the environment in terms of quantity in Bq, taking into account the specific nuclides, release conditions, and so on.

(2) The radiation exposure dose to the public in the vicinity of the plant in Sv for an individual or the collective dose in person-Sv for the total population or both, taking into consideration the meteorological factors for atmospheric dispersion and the population distribution.

(3) The potential number of fatalities or the fatality rate due to stochastic and/or nonstochastic radiation influences or both.

(4) The property damage.

Using PSA for severe accidents at nuclear power plants, the probability of release of the radioactive materials in expression (1) and the probability of radiation exposure in (2) can be obtained. These probabilities can be analytically estimated using reliability data for the equipment and systems concerned, but the analyses should use appropriate evaluation methods for the atmospheric dispersion and radiation dose calculation. Probabilities determined using these techniques are called "engineering probabilities" in this paper.

Theoretically it is possible to estimate the number of fatalities related to potential accidents mentioned in (3) by multiplying the dose or the collective dose in (2) by the probability of the attributable death which is similar to the fatality probability coefficient or the dose rate effectiveness factor. Fatality probability coefficients have been obtained from studies by medical experts of the ICRP, the UNSCEAR, etc. Although these studies involved data mainly from patients actually exposoured to radiation, conservative and controversial assumptions and uncertainties exist in these coefficients. However, if an accident actually occurs, such as the Chernobyl accident, and if the exposure dose to the public is able to be estimated, the fatality probability coefficient is available to estimate the number of potential fatalities for the entire population affected by the accident.

In addition, theoretically it may be possible to evaluate the risk of fatality related to potential accidents by multiplying the results obtained in (3) by the "engineering probabilities" of incurring the dose or the collective dose resulted from exposure discussed in (2). However, the "engineering probabilities" and dose rate effectiveness factors have many uncertainties, engineering assumptions and judgments. When considering the different elements which exist, it is not logical to estimate the risk of fatality related to potential accidents by multiplying the "engineering probability" by the dose rate effectiveness factor. This method would mislead the general public into the belief that the fatality risk estimated is the actual risk the nuclear power plant introduces. Care should be exercised in the ICRP 1990 Recommendation (draft) when these issues are mentioned, for example in section 5.6 (potential exposure situations). It is recommended that due to the present state of the art, the health risk matters discussed in the radiation protection policy should be independent from the PSA in the nuclear safety policy, which relates mainly to engineering problems, particularly the specific topics of low probability incidents in nuclear power plants. Additionally, it is recommended that the adverse effects of nuclear accidents shown in the safety goal should be expressed as mentioned in (1) or (2).

There are no official statements which express the safety goal of nuclear power plants in Japan in terms of fatality rate or fatality risk.

The safety goal in Japan is expressed using the principal philosophies derived from the codes and guidelines – the siting guideline, the general design guideline, the guideline on safety evaluation of design in connection with unusual incidents and accidents, and others. As stated before, the basic safety policy is to prevent prompt fatalities in the public and workers and to mitigate the societal radiational risks as much as possible if such a highly unlikely accident were to occur.

Figure 1 shows the safety goal proposed by the author, which incorporates the safety policy in Japan as described above. The safety goal curve indicates the limit value of a whole-body dose, C Sv, to a model off-site individual per incident or accident as a function of the probability of occurrence of an incident or an accident per reactor-year, P. Severe, highly improbable meteorological conditions are statistically assumed. The probability is defined as the maximum probable incurred radiation exposure dose to an off-site individual resulting from an accident.

The significant characteristics of this safety goal curve are:
(1) Normal operation target, 0.05 mSv vs. $1 - 10^{-1}$ per reactor-year.
(2) Maximum permissible dose limit for individual, 5 mSv vs. 10^{-2} per reactor-year (less than once in a reactor lifetime).
(3) Maximum permissible dose limit for a worker, 50 mSv vs. 10^{-4} per reactor-year (less than once in the life of all LWRs in Japan).
(4) Reference dose limit for the siting evaluation accident, 0.25 Sv vs. 10^{-7} per reactor-year.
(5) Anticipating that emergency procedures would keep the exposure dose less than 0.25 Sv, even in an accident more severe than the siting evaluation accident, a convex curve is shown for probabilities less than 10^{-6}. It is thought that the 0.25 Sv proposed by the Radiation Protection Commission as a reference dose value for the dose limit for an accident can be maintained by evacuation.
(6) Even if the emergency procedures discussed in (5) are neglected, the exposure dose does not exceed 1 Sv for the range of reasonable probabilities. This is shown by the logarithmic (straight line) portion of the curve for probabilities less than 10^{-6}.

The operating experience of nuclear power plants in Japan demonstrates they have to date met this safety goal and the prescribed limits in the radiation protection policies. Potential severe accidents should be evaluated from a Level 2 PSA to determine if nuclear power plants in Japan would be able to meet the safety goal associated with various accidents. However, it must be emphasized that achieving the safety goal can only be assured by accumulating day-to-day plant performance which is safe and reliable.

8. Conclusions

The basic policy for the nuclear safety in Japan is that top priorities are given to measures to prevent abnormal events rather than to those which mitigate the consequences of abnormal events and accidents. Abnormal events are prevented by design, manufacture and construction using high technology, careful quality assurance and well trained and reliable operators and managers. The basis of safe design, a worldwide concept, is the use of a highly conservative design which incorporates defense-in-depth, multiple barriers, single failure criterion, etc. Measures to preserve the integrity of systems, such as detecting slight defects or their symptoms at an early stage and properly exchanging or repairing the defective component, are very important in operational maintenance and management. Nuclear power stations in Japan are recognized as highly safe and reliable by the general public due to the twenty-four-year accumulation of operating performance. The TMI-2 and Chernobyl accidents taught us that a reactor accident will have worldwide social and political influence as well as technical one. To achieve further improvement in safety and obtain proper understanding by the public, it is important to advance deliberate research and reviews on PSA, and on severe accidents and countermeasures for them.

Reactor facility troubles which occurred in the early developing period seem to have been resolved with technologies accumulated for nearly 30 years. Now and in the future, research and investigation on aging degradation will be important.

The quantitative interpretation of PSA and the safety goal should be carefully considered, taking into account their applicable constraints. The nuclear safety issues related to PSA and the safety goal should be discussed now to obtain international consensus on fundamental safety policies to harmonize rational safety concepts.

It is believed that the cumulative experiences in Japan can contribute to the further improvement of safety of nuclear power plants throughout the world. Thus, the international exchange of information should be stimulated.

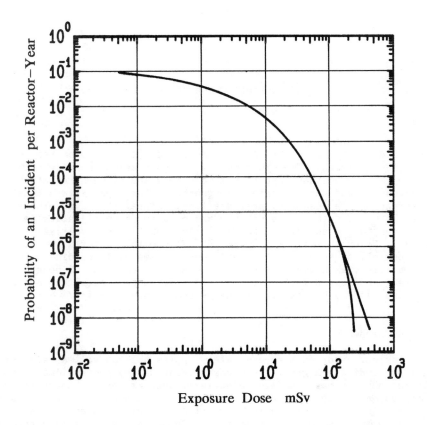

Figure 1   Proposed Safety Goal

DISCUSSION

R.H. CLARKE, United Kingdom

Lets start with the discussion on the three safety papers.

A. GONZALEZ, IAEA

The paper presented by Dr. Ryder suggests different approaches to safety among European countries. The other papers in the session indicate differences between European and overseas approaches to safety and between radiation protection approaches and nuclear safety approaches. Some specialists consider these differences of a cosmetic, presentational nature; but some others including myself believe that some of them are of a fundamental nature (e.g. the treatment given to the so-called "societal risk" or "collective risk" or "detriment"). Given this situation, do you believe that experts have a professional responsibility to search for consensus on some basic guiding principles on radiation safety as a way to facilitate the continuous use of beneficial practices (e.g. nuclear power) involving radiation exposure (either actual or potential) and, if so, how can we discharge such responsibility?

L.G. HÖGBERG, Sweden

I think our paper on European safety approaches, as well as some other presentations, indicate a growing consensus among safety professionals on basic safety philosophy and basic safety objectives, such as the 10^{-5} per reactor year safety objective for core damage frequency, or the still lower probability for off-site releases, requiring substantial emergency measures such as long-term evacuation and land usage limitations. These safety objectives and the ways to achieve them are broadly in agreement with the INSAG-3 report on basic safety principles. However, this growing consensus in the professional safety community does not seem to be matched by a similar international convergence among political decision-makers, something that may be crucial to the future of nuclear power, taking into account the importance of questions like: "Are my neighbour's reactors safe enough?"

F.J.REMICK, United States

There is no question that nuclear professionals should work together to seek common standards for the protection of radiation workers and the general public. However, I foresee the potential for agreement on general principles, but some variation in details and to modes of implementation. Although ideally preferable, I do not see exact, identical standards as a prerequisite to the future of nuclear power utilisation.

H. UCHIDA, Japan

We all have common objectives of assuring radiation protection and nuclear safety. Therefore, as a baseline argument, I personally believe we should patiently search for a way to reach a worldwide consensus on this issue. Especially it is recommended that an international framework be established for basic safety principles. The activities of INSAG have contributed significantly to this point with respect to nuclear safety. It is noted, however, that these are rather fundamental and qualitative in nature, hence they must be further developed or elaborated to meet standards requested in each nation by carefully and prudently considering their own regulatory, social and political systems.

Regarding the radiation protection and its interface with nuclear safety, the recommendations of ICRP are to be respected. I see no problem in applying them in the case of the normal exposure resulting from the operation of nuclear installations. However, it seems that the treatment of "potential exposures" is still controversial. I believe, therefore, that its actual implementation in nuclear safety practices must be done with extreme care.

M. LAVERIE, France

En matière d'uniformisation réglementaire, il convient de distinguer :

– **Les objectifs de sûreté** : Même si le contexte politique et historique a induit des différences, qui sont d'ailleurs souvent plus de forme que de fond, l'uniformisation progresse pour la formulation des objectifs de sûreté. Notamment dans le cadre des organismes internationaux (AIEA, AEN, CEE), le consensus se construit et l'on imagine mal que les différents pays puissent exprimer des objectifs sensiblement différents pour la sûreté de leur prochaine génération de réacteurs.

– **Les modalités pour atteindre les objectifs** : Elles sont effectivement diverses. Aucun processus de rapprochement ne peut se produire si les exploitants et constructeurs de chaque pays continuent à soumettre à l'approbation de leurs autorités de sûreté respectives des pratiques différentes.

Ce n'est donc que dans la mesure où des modalités techniques seront soumises par un acteur industriel aux autorités de sûreté de plusieurs pays qu'une uniformisation des positions réglementaires pourra être recherchée. Il est clair que, dans un tel cas, les autorités de sûreté concernées devront tout faire pour définir entre elles des positions cohérentes.

Mais ce processus ne peut qu'être postérieur à la naissance de projets industriels ayant un caractère international.

R.J. BERRY, United Kingdom

I would like to particularise Dr. Gonzalez's plea; Dr. Ryder's presentation on common European safety goals identified "harmonization on the highest standards". Admirable though this sounds, it is a recipe for disaster in that individual, national authorities may well set <u>unreasonably</u> high standards for these parts of the nuclear industry which they do not themselves have to

authorise within their borders. For example, Prof. Birkhofer mentioned in
response to me this morning a target for dose to most exposed individual from
Germany nuclear power stations of 10 µSv per year. Our reprocessing plant at
Sellafield could not meet this goal for tens of years because of design decisions taken more than 20 years ago and even if we ceased all current discharges
tomorrow. Yet we perform well against a UK target of 500 µSv to the critical
group where for last year the achievement was 170 µSv. Harmonization must be
based on "safe enough" if a viable European nuclear industry is to survive.

E.A. RYDER, United Kingdom

I am grateful to Dr. Berry for making the point and for giving me notice
of it. The text of our paper, and my presentation of it, both carefully used
the words: "If compromise is to be reached it should be in the direction of
the highest agreed standards."

Dr. Berry's point illustrates how the debate will go; he represents a
fuel processing company, an operator. The authors of our paper are all
regulators and, since the public is not present at these gatherings, we
regulators have to do our best to represent their interests.

A. BIRKHOFER, Germany

Comment to Dr. Berry. Our radiation protection regulations state
300 µSv per year as a maximum dose to an individual of the general public. The
actual doses, however, to the surrounding population of Germany nuclear power
plants are of the order of 10 µSv per year.

L.G. HÖGBERG, Sweden

I think that in the past we have seen too many bad examples on "international harmonization" based on the least common denominator instead of the
best available technology or practice, also proven to be commercially viable.
As a regulator, with the protection of the public as my prime responsibility, I
cannot defend regulatory decisions which do not take into account the best
available technology, especially if it is proven to be commercially viable, and
in particular with respect to new installations and practices.

P.Y. TANGUY, France

La question essentielle posée par une harmonisation internationale des
règles de sûreté me paraît être la suivante : à partir du moment où il existe
un consensus sur les objectifs, comment obtenir qu'un projet acceptable dans un
pays ne le soit pas dans un autre, parce qu'il ne satisfait pas les règles en
usage, alors que son niveau de sûreté devrait le rendre acceptable ? M. Lavérie
a suggéré une action de l'industrie. Elle est en cours. Je pense que les autorités de sûreté devraient évaluer la valeur de leurs règles en terme de sûreté,
à la lumière des Analyses Probabilistes de Sûreté. Ainsi un dialogue pourrait
s'établir sur le risque et pas seulement sur la conformité aux règles.

A.P.U. VUORINEN, Finland

It is natural that the need to develop more or less consistent safety goals and safety approaches is seen differently in large and small countries. Experts, politicians and active members of the public are looking beyond the borders in small countries and the situation which is seen is rather confusing. At the time when public acceptance is a razor's edge, the current situation is not favourable to the nuclear community.

R. NAEGELIN, Switzerland

According to the paper by Commissioner Remick, the NRC does not want a "large release" criterion which would be more restrictive than the health objectives of the safety goals (p.5). However, results of PRAs and the experience of Chernobyl indicate that the health effects of severe accidents are comparable or even lower than such effects of other catastrophes, while the contamination of the environment and the resulting loss of land may go beyond corresponding effects of catastrophes experienced before. Correspondingly one would expect that the most restrictive criterion or safety goal should be connected with those other effects.

F.J. REMICK, United States

Admittedly, our Safety Goals, do not include an objective on limiting land contamination in case of an accident. This was debated during the development of the Safety Goals, but not included when it was decided to focus on public health and safety, the mandate of the NRC.

F.J. TURVEY, Ireland

How can one be sure that there is consistency between one PRA and another? Presumably in the U.S. one must employ many teams to perform PRAs and this must introduce inconsistencies. If this is so, how can one justify PRA playing a useful role in the setting of safety goals?

F.J. REMICK, United States

Many PRAs have been performed in the United States; however, admittedly, only a small subset are state of the art, level-3 PRAs, which include both internal and external initiators. As more state of the art level-3 PRAs are available, it is our view that these will provide adequate information to make an engineering judgement as to whether or not the NRC's regulations are producing nuclear power plants which as an ensemble meet the quantitative health objectives and thus the qualitative safety goals defined by the NRC.

R.H. CLARKE, United Kingdom

You have explained that the safety goals will be measured against the average of the results of PRAs over a number of reactors. What will you do about individual reactors which show higher levels of "risk" than your safety goals? That's the most probable source of an accident.

F.J. REMICK, United States

When we have available a sufficient number of "state of the art",
level-3 PRAs which include both internal and external initiators, we will compare these with the quantitative health objectives of our Safety Goals. From
this we will judge whether or not our regulations are doing an adequate job of
providing an ensemble of plants that appear to meet our goals. If a particular
plant is found to represent an outlier, consideration of modifications would be
made under our Backfit Rule (10 CFR, Part.50.109).

A. BIRKHOFER, Germany

Advanced reactor concepts will include as much as possible accident
management and consequence mitigation actions. I will foresee for the future
with new reactors designs that off-site emergency measures should not become a
prequisite for licensing. The source terms to be considered will be much
smaller than the ones presently considered. Could you see this future also in
the USA for advanced reactors?

F.J. REMICK, United States

One industry-prescribed criterion for advanced reactors in the United
States is that the public risk of accidents should be so low that emergency
planning zones could be reduced to that of the plant site. Without prejudging
whether this should be permitted, I must admit that in the political climate
and public concern evidenced in the United States I foresee considerable difficulty in the Commision making such a regulatory change at this time.

P.Y. TANGUY, France

Le Commissaire Remick indique dans sa communication que l'objectif
général de limiter la probabilité de rejets radioactifs importants ("large
release") à moins de 1 sur 1.000.000 par an peut être facilement compris par le
public. Peut-il nous donner des précisions sur l'accueil du public ? En Europe,
je crains qu'un tel objectif soit simplement interprété comme : un Tchernobyl a
une probabilité de 10^{-6} !

F.J. REMICK, United States

My reference to attempts to gain public understanding and acceptance
applied to the Quantitative Health Objectives which were stated in terms
intended to be readily understood by the public. I believe that time has shown
that these objectives have been understood and accepted as an expression of
"how safe is safe enough" from the operation of nuclear power plants in the
U.S. As we have not yet defined what we mean by a "large release" in our Large
Release Guidelines, I do not claim public understanding and acceptance, in that
respect.

A. GONZALEZ, IAEA

I am afraid that I disagree with Mr. Högberg concerning safety approaches that either there are no conceptual differences or that they are converging. Both the papers and the subsequent discussion show the existence of conceptual differences.

I do agree, however, with Mr. Lavérie that there is no (or little) room for consensus at the engineering safety solutions; but I'd like to insist in my implicit proposal that it would be desirable to obtain a consensus at the conceptual level on fundamental safety principles. This consensus is not necessary for public relations purposes as suggested by Mr. Vuorinen but rather for pure technical reasons.

E.A. RYDER, United Kingdom

Can I comment on Dr. Birkhofer's question to Mr. Remick? The UK takes a prudent view. It believes that accidents can happen and that there must be emergency arrangements to cater for them. The UK requires detailed emergency plans to cover about 5 km radius round each nuclear plant. These are capable of extension, should it be necessary.

The Sizewell B public inquiry required and endorsed a table-top demonstration of the extendibility of the emergency plan out to about 10 km, based on a hypothetical large release of radioactivity.

The subsequent public inquiry for Hinkley Point C also endorsed this approach of a detailed plan based on a reference accident supplemented by the capability of extension to cope with a hypothetical large release similar to that used for Sizewell B. The Inspector further recommended that this should be used for all nuclear power stations. The Secretary of State for Energy and HM Nuclear Installations Inspectorate of HSE have accepted his formal recommendations on this point. In summary, we don't see that we can drop emergency planning.

Mr. Abel Gonzalez asked whether a nuclear renaissance depended upon reaching international agreement on safety goals and radiation protection requirements.

I am interested to see how a nuclear renaissance might happen. I doubt if a unified theory of radiation protection and nuclear safety philosophies will be available on the timescale in which some countries need to decide on more generating plant, possibly nuclear power plant. I doubt too that such a unified theory is necessary, even if it is desirable.

There may well be some increased convergence of safety goals, at the highest level, but I suggest that some form of greater agreement on engineering judgements on design details is more likely as a practical way forward. It will be essential for nuclear safety regulators to be involved in any discussions between manufacturers or utilities, of course.

F.J. TURVEY, Ireland

In the paper entitled "European Safety Approaches" the authors say "Certainly, actual levels of achieved safety appear to be consistent across Europe". This implies that in Europe we have found a method of measuring reactor safety. Is this really so?

J.M. REED, United Kingdom

Questions of safety are discussed within a CEC Working Group. In those discussions many comparisons of safety approaches have been studied, not only probabilistic but deterministic approaches. It is acknowledged that there are differences of approach, and no methods of actually measuring safety, but the consistent results of studies have shown convergence of views and agreement that methods of achieving safety by engineering solutions are broadly similar. Achieved levels of safety, therefore, certainly at the deterministic safety standards level, appear consistent.

H. KOUTS, United States

I believe that the move toward unification of safety objectives will take place naturally, because of the importance of safety across national boundaries as Dr. Vuorinen has said. It will perhaps take place, as Dr. Tanguy has suggested, through the moves industry has been taking.

R.H. CLARKE, United Kingdom

Let's now have some discussion on the radiation protection philosophy as presented by Dr. Beninson.

F. COGNE, France

A partir des mêmes données biologiques, la CIPR arrive à deux valeurs limites différentes pour le radon d'une part, pour l'énergie nucléaire d'autre part. Ceci est particulièrement surprenant et paraît en tout incohérent.

P.Y. TANGUY, France

Je reprendrai la question de M. Cogné sous une autre forme. Comment peut-on expliquer aux habitants de Biélorussie affectés par les retombées radioactives de Tchernobyl, qu'à partir des mêmes données biologiques, on ne leur applique pas les mêmes limites que celles fixées par la CIPR pour la population autour des centrales (1 mSv par an, soit 70 mSv dose totale vie) mais plutôt une limite de 350 mSv ?

D. BENINSON, ICRP

There are not two dose limits. One refers to an action or intervention level, the other to a dose limit. Mr. Cogné referred on the one side to radon, a "natural component", and on the other side to consequences of the use of nuclear energy, a man made component. These things are different. Let's take a more extreme case than radon, the cosmic radiation. If the cosmic radiation had been much higher than what it is at present, what could we do to protect ourselves? Not much I am afraid. On the contrary if we introduce a practice (justified by the benefits) and we can adapt it to the standard of safety that we want to have, why shouldn't we do that if the costs are reasonable. Thus the explanation is the different nature of a limit and an intervention level. In the first case the increase of exposures (justified by the benefits from the practice) is regulated by dose limits. In the second case an existing exposure situation is examined to see if a remedial action to reduce doses would do more good than harm taking into account the risk of the remedial action itself and the social cost. Let me illustrate this. Nobody would probably take a remedial action against an existing situation causing 1 mSv, but would certainly protect against a man made source causing 1 mSv, if the costs were reasonable. These two cases have to be treated differently. The main reason for the confusion between the two cases is, I believe, a "gut" feeling of threshold when talking limits, which of course is wrong. The basic idea of linearity, non-threshold linearity, is not deeply accepted. Many people feel that there is a threshold at the dose limit and thus think that if you are below the limit there will be no problem. This is the difficult point, I think. If one goes away from this false concept, it will become clear that at the same risk in one circumstance you would do something, in another you would not. If your remedial action, for example, would cause more harm than that you want to get rid of, you would not just do it.

Concerning the situation in the Soviet Union I would like to say that I have questioned the wisdom of basing an action level (350 mSv total life projected dose) on a dose limit (5 mSv in a year, old ICRP value). There is no justification of doing that.

A.P.U. VUORINEN, Finland

Dr. Beninson, you criticized 5 mSv and 350 mSv limits used in the Soviet Union and mentioned also that nobody would take any actions due to 1 mSv of dose. I would not be so sure about that. We all remember the questionable intervention levels for food introduced in Europe and elsewhere after Chernobyl. There is little hope that the actions next time (if that ever comes) would be fully rational. To be honest at least part of the hysteria now so deep in effected areas in the Soviet Union, is of western origin.

D. BENINSON, ICRP

I agree that the action levels established in the West for food control were not reasonable at least in the beginning.

G.N. KELLY, CEC

The logic described by Dr. Beninson on the differences between dose limits for controlled practices and levels for intervention is impeccable. However, despite the clear conceptual differences, misunderstandings remain even among those practising in the radiation protection field. More needs to be done in communicating effectively these concepts to both the radiation protection and nuclear safety communities.

R.M. DUNCAN, Canada

Regarding Dr. Beninson's comments on misconceptions which exist about the functions of dose and risk limits, we have to recognise that we ourselves contribute to these misconceptions by explaining dose limitation in terms of "acceptable", "not unacceptable" and "unacceptable" regimes of dose or risk, with the dose limit fitting between the two latter regimes. The public and workers are not likely to feel more comfortable with this terminology than with the use of "safe" and "dangerous" to describe the two regimes on either side of the dose limit.

Similarly, by referring to some "percent of the dose limit" as a measure of how effective dose or risk control is in a nuclear facility, we leave the impression that use of the dose limit is the only effective way of controlling exposures.

Dr. Beninson responded to Dr. Ilari's question about the necessity for the ICRP to recommend lower dose limits by referring to the case of uranium mines. However, the impact on other facilities is also quite real, because we are actually looking at a change in risk coefficient and this affects all facilities at all levels of exposure or release, granted to a lesser degree the further the effects of these facilities are from the dose limit. So, in reality, the pressure will be on by public and workers to have all facilities reduce their impacts in some manner relative to this change in risk coefficient.

E. GONZALEZ GOMEZ, Spain

I am afraid that with the actual ICRP recommendations, that establish not only risk estimates but also limits, the ICRP gets away from its scientific stand into the decision making process, leaving little room for taking decisions in accordance with the social factors of the different areas, situations and nations in the world.

Session 3

ACHIEVEMENT OF RADIATION PROTECTION
AND NUCLEAR SAFETY OBJECTIVES
IN THE REGULATORY AND TECHNICAL PRACTICE

Séance 3

REALISATION DES OBJECTIFS DE LA RADIOPROTECTION
ET DE LA SURETE NUCLEAIRE DANS LA PRATIQUE
AU PLAN REGLEMENTAIRE ET TECHNIQUE

Chairman – Président

D. BENINSON
(ICRP)

L'APPLICATION DES PRINCIPES DE LA CIPR EN VUE DE LA MAITRISE DES RISQUES DUS A L'EXPOSITION POTENTIELLE

R. Cunningham[1]
U.S. Nuclear Regulatory Commission
Washington, Etats-Unis

A.J. González[1]
Agence Internationale de l'Energie Atomique
Vienne, Autriche

RESUME

Jusqu'ici, les recommandations de la CIPR visant la radioprotection s'appliquent principalement aux radioexpositions dont il est quasi certain qu'elles acconpagneront le fonctionnement normal des sources de rayonnements. Il est prévu que les nouvelles recommandations de la CIPR traiteront de façon plus exhaustive la sûreté radiologique en prenant en considération l'exposition susceptible ou non de se produire, mais à laquelle il est possible d'assigner une probabilité de survenue (exposition potentielle). Dans la présente communication, les auteurs examinent les problèmes et les principes relatifs à un système de sûreté radiologique qui tient compte des normes tant de radioprotection que de sûreté nucléaire, et qui couvre les expositions tant normales que potentielles. Pour énoncer ces principes, on a recours à une interprétation et une extrapolation des principes de justification, d'optimisation et de limitation de dose actuellement en usage pour l'exposition normale.

---

(1) Les opinions exprimées par les auteurs se rapportent à leur participation à un Groupe de travail de la CIPR chargé d'examiner l'exposition potentielle et ne reflètent pas nécessairement les positions des organismes dont ils sont les agents.

THE USE OF THE ICRP PRINCIPLES FOR
CONTROLLING RISKS FROM POTENTIAL EXPOSURE

R. Cunningham(1)
U.S. Nuclear Regulatory Commission
Washington, Unites States

A.J. González[1]
International Atomic Energy Agency
Vienna, Austria

ABSTRACT

Heretofore, ICRP recommendations for radiation protection mainly apply to radiation exposures that are expected to occur with near certainty during normal operation of radiation sources. It is anticipated that the new ICRP recommendations will deal more comprehensively with radiation safety by including consideration of exposure which might or might not occur, but for which a probability of occurrence can be assigned (potential exposure). This paper discusses issues and principles for a system of radiation safety which accommodates both radiation protection and nuclear safety standards and covers both normal and potential exposures. The principles are formulated by interpreting and extrapolating the principles of justification, optimization and dose limitation currently employed for normal exposure.

---

(1)   The views expressed by the authors are related to their participation on an ICRP Task Group examining potential exposure and do not necessarily represent the views of the Agencies in which they are employed.

1. INTRODUCTION

The ultimate aim of both radiation protection and nuclear safety is to adequately protect people from the harmful effects of radiation. Although this is a common objective, the corresponding standards (and the manner in which they are expressed) and assessment methodologies (and the requisite skills to employ them) vary considerably, depending on the specific issue being addressed. Because of this, the thread of logic tracing through to the end points of assuring and demonstrating an appropriate level of radiation safety is often lost. It can result in fragmented approaches and imbalances in the overall degree of safety provided. The International Commission on Radiological Protection (ICRP) is currently revising its basic recommendations for radiation protection. One objective of the revision is to treat radiation safety more comprehensively and coherently in its system of protection.

The system of protection recommended by ICRP in ICRP Publication 26 mainly applies to radiation exposures that are expected to occur with near certainty during the normal use or operation of radiation sources. This type of exposure is referred to as "normal exposure" in a draft revision of the Commission's recommendations which has been widely circulated within the radiation protection and nuclear safety communities.

It seems that the new ICRP recommendations will deal more comprehensively with radiation safety by including consideration of exposure situations which might or might not occur, but for which a probability of occurrence can be assigned to the actual occurrence of the exposure of people. Such situations include accidents and other abnormal events which can lead to radiation exposure in addition to the normal exposure resulting from the use of a radiation source. This type of exposure is termed "potential exposure" in the ICRP draft revision.

There is a third type of situation which the new recommendations would also address in its more comprehensive system of radiation safety. It concerns the protection of people from radiation sources that are not deliberately introduced, but which pre-exist in the environment. These can be sources found in nature or result from unplanned situations, e.g., large scale contamination resulting from a major reactor accident. The approach to protection in this type of de facto situation is distinctly different from protection against "added" normal or potential exposure in that protection involves some kind of intervention intended to "subtract" part of the exposures which are being incurred or expected to be incurred. This aspect of a more comprehensive system of radiation safety is mentioned for completeness, but will not be discussed further in this paper.

The purpose of the paper is to suggest the principles for a consistent and coherent system for radiation safety in general which accommodates both radiation protection and nuclear safety standards, and covers both normal and potential exposure situations. These principles have been formulated by interpreting and extrapolating the principles of justification, optimization and dose limitation currently employed for normal exposure. The ultimate goal of the system would be to exercise restraint upon the radiation harm that may result from potential conditions or events, which occurrence probability can be anticipated, as well as that due to normal exposure.

It is recognized that while it is possible and may be desirable to conceptually incorporate potential exposure situations in any system of radiation safety, the methods by which potential exposures can be assessed, managed and ultimately controlled in practice must undergo an evolutionary process. It involves many complex issues, a number of which will be discussed in this paper. Therefore, a second purpose of presenting the conceptual basis for incorporating potential exposure in the system for radiation safety at this seminar is to stimulate discussion about how it can be applied in practice with useful results.

2. APPLICATION OF BASIC RADIATION PROTECTION CONCEPTS TO POTENTIAL EXPOSURE SITUATIONS

Two basic concepts used in formulating radiation safety recommendations or standards for normal exposure situations must be treated differently for potential exposure situations. They are risk and detriment.

The ICRP has used the term risk to indicate the total individual probability of attributable death from radiation effects. Such a probability can be defined as the product of the probability of incurring a dose and the lifetime conditional probability of attributable death due to the dose if it is actually incurred. For normal exposure, the first probability is assumed to be unity and the doses sufficiently small as to postulate a linear relationship between dose and the probability of death. The risk then is directly proportional to the dose. For potential exposure, the probability of incurring a dose is less than unity and the dose, if incurred, could be sufficiently high as to produce deterministic (non-stochastic) effects. Therefore, the risk from potential exposure is neither dependant on the potential dose alone nor necessarily linearly proportional to the dose, but a convolution of two probabilities: the probability of incurring a dose and the probability of attributable death given the dose.

Detriment is used by the ICRP to mean the overall, expected (statistical) harm that would be experienced by a population group and its descendents as a result of exposure from a radiation source. In the case of normal exposure, detriment is assumed to be linearly proportional to the collective dose because each individual risk is proportional to the dose incurred by the individual and the sum of all risks must be equal to the mathematical expectation of harm. The specification of detriment for potential exposure is more complex for two fundamental reasons: (i) doses delivered when an event occurs could be large enough to result in deterministic effects in some individuals and, therefore, in such instances the use of collective dose to reflect expected harm to the exposed group would not be appropriate; and (ii) using the product of the probability of an event and some quantitative expression of consequences should it occur – the expectation value of the consequences – conceals the fact that the outcome will be either no consequence if the event does not occur or the full consequence if it does. (There is not, in itself, reciprocity between reduction in the probability of an event and reductions in the scale of consequences and if not properly taken into account, assessment can lead to a conclusion that a high probability event with minor consequences and a very low probability event with major consequences are equally detrimental if the expectation values of the consequences are the same).

Taking these complications into account, it seems that a more comprehensive approach should be adopted for specifying detriment for potential exposure situations. One possibility is that of multi-attribute specification. The detrimental factors (or attributes) attributed to the available safety options could be identified and quantified. They can then be given a weighting factor judged to represent its importance. The weighted attributes of a safety option can then be aggregated to provide a figure of merit (or demerit) which can be compared, as a whole or individually, with the weighted attributes of other options. Either method leads to a partially quantitative and partially qualitative basis for assigning detriments for alternative safety options. (A simpler approach to specifying detriment is possible if anticipated doses are expected to be small if the event occurs. In such instances it is acceptable to use the product of the expected dose and its probability of occurrence as if it were a dose that was certain to occur).

3.   UNCERTAINTIES IN ASSESSMENT OF NORMAL AND POTENTIAL EXPOSURE

The perception exists that assessments of risks and detriments in "normal" exposure situations are always subject to smaller uncertainties than those in "potential" exposure situations. This is not necessarily the case. For example, dose assessments for normal exposure due to intakes of radioactive materials can involve quite significant uncertainties because of factors such as the difficulties in the quantification of the intake and the physiological differences between the person actually contaminated and the standard person for whom the metabolic models were established. Also, while the uncertainties about the risk of high doses are relatively low, risk estimates for normal exposures with low doses and dose rates are uncertain owing to lack of quantitative data on the dose-risk relationship in this range. The assessment of individual risk due to potential exposure situations is subject to additional uncertainties, such as those associated with the exposure probability. But, in some instances, these will be small compared with the uncertainties associated with the risk due to the dose received (e.g. for relatively high probability/low consequence events). In other instances, however, the uncertainties associated with the probability of the event can be so large as to make the assessment meaningless (this would typically be the case for very low probability events). The problem of uncertainties, therefore, is common to both normal and potential exposure situations and should be taken into account in both cases.

4.   PRINCIPLES FOR LIMITING RISK FOR NORMAL AND POTENTIAL EXPOSURE

The ultimate objective of a system for radiation safety is to exercise restraint on radiation risks and detriments to people from all exposure situations, both normal or potential, that can be anticipated to occur from the introduction of a source of ionizing radiation. Therefore, normal and potential modes should be considered in a comprehensive and coherent protection system. This can be accomplished by including consideration of potential exposure in the three principles of the system of protection presently recommended by ICRP. These principles can be stated briefly as follows:

(a) The introduction of a practice involving radiation exposure shall deliver more benefit than harm (or <u>justification of a practice</u>).

(b) The safety level for each radiation source shall be the best under the prevailing social and economical circumstances (or <u>optimization of safety</u>).

(c) The probability of individuals or their descendents to incur severe radiation harm shall, under any circumstances, be constrained by preselected limits (or <u>individual risk limitation</u>).

4.1 Justification of a Practice

This requirement simply specifies that, in order to permit the introduction of a practice (which is expected <u>to add</u> radiation exposure and, consequently, radiation risk to people) more benefit than detriment should be expected. The implications of this requirement have been analyzed in terms of simple cost-benefit analysis by the ICRP and other organizations, under the limited framework of normal exposure situations. When moving from normal exposures to the wider scope of potential exposure situations, the practical application of the principle of justification becomes more complicated. For some potential exposure scenarios, the probability of occurrence can be very low but, if the scenario really happened, the consequences could be 'unjustifiably' high. It is not clear how these situations should be included in an assessment of justification and whether this requirement should be applied in a prescriptive manner. Rather, it seems that it should form part of a system's general declaration of principles, with the only implication of stressing the ability of authorities to exclude the introduction of a practice on the basis of justification arguments.

4.2 Optimization of Safety

The ultimate level of safety applied to a radiation source results from a choice among feasible alternative safety options. A relevant role for those making decisions should be to satisfy themselves that the best safety option under the prevailing circumstances has been selected. The conceptual process governing the selection of safety options is termed "<u>optimization of safety</u>" and its aim is to keep radiation risks and detriments arising from the source to levels deemed to be as low as reasonably achievable, after taking into account the social and economic factors constraining the selection.

The first step towards safety optimization is the identification of the relevant factors to be taken into account in the process. For normal exposure situations the factors considered in international guides are the <u>radiation detriment</u> (usually expressed as proportional to the collective dose) and the protection efforts (sometimes expressed in terms of the protection costs). The treatment of potential exposure situations will necessarily involve other factors such as the probability of the exposure, the distribution of risks, and the total consequences should the exposure occur.

Additionally, potential exposure situations can result in doses producing deterministic, clinically observable effects which can be traced from the affected individual to the exposure, and it is necessary to consider whether or not such doses should be weighted differently than low doses giving rise to effects of a statistical nature only. For a variety of reasons, such as the different manifestation in time of the effects and the different perceptions of the risk involved, it is possible that a radiation safety system should apply progressively higher weighting to the effects attributable to high doses. A possible problem resulting from different dose weighting is the shifting of risks and detriments from potential exposure to normal exposure or from public exposure to occupational exposure and vice versa. For instance, occupational risk due to backfitting measures at nuclear power reactors, which are intended to reduce the public risk from accidents, may exceed the corresponding benefits from reduction in potential exposures of the public.

Final decisions about optimized safety may result from an integration of considerations for both normal and potential exposure situations. Decisions will very likely involve both quantitative and qualitative components. Since the factors involved can have different dimensions and available options may have multiple attributes with no easily assessed bottom line, a decision-aiding process should normally be used to decide on the optimum safety option. From the available techniques, the process termed multi-attribute utility analysis appears the most appropriate for accommodating both normal and potential exposure situations with quantitative and qualitative components.

4.3   Individual Risk Limitation

An optimum safety option arrived at on the basis of an unconstrained optimization process will not be necessarily acceptable unless it meets the requirement that no individual shall be subjected to a probability of radiation harm greater than a pre-established level. Thus, the optimization process should be constrained by individual risk limits which should be established prior to the optimization process.

For normal exposure situations, the probability of radiation harm to individuals is constrained through the imposition of individual dose limits. Assuring that doses do not exceed limits for normal operations is accomplished in a variety of ways which are broadly encompassed by pre-operational planning, assessment and design, and monitoring safety system functions, dose-rate levels and dose accumulation as operations progress.

For potential exposure situations the limitation of the individual probability of radiation harm is complex. To provide consistency with the system of dose limitation for normal exposure, it seems logical that limits of risk from potential exposure be of the same order of magnitude as the risk implied by the current dose limits. To maintain risk within overall limits or protection objectives, a safety system should establish practical constraints for specific sources and for specific potential exposure scenarios or classes of events leading to exposure.

Constraint of risks from potential exposure situations can be obtained by constraining the probability as well as the magnitude of the exposure; i.e. by limiting the probabilities of doses being incurred and/or the doses themselves. This may be achieved through measures such as assuring the reliability of systems, establishing operational procedures, and controlling human actions. These measures should be identified in the initial assessment as being essential in attaining the confidence that unacceptable individual risks will not be imposed. Additionally, technical and administrative measures to further reduce the potential doses should also be taken into account: factors determining the effectiveness of mitigation features, such as containment of radioactive material, and emergency measures may have to be determined on a case-by-case basis in a prescriptive way.

One procedure for applying individual-related constraints to probabilistic events is to express probability limits as a function of the dose that will be delivered should the event actually occur. Such a limit will express the maximum probability that can be permitted for an estimated dose from event sequences leading to exposure. Assessments involve the probabilities of incurring doses by taking into account various possible scenarios or classes of events, and the reliability of the relevant safety systems and procedures. These probabilities can then be compared with those established by authorities for various potential doses. Should the comparison be unfavourable or not satisfactory, the overall reliability of safety systems would need to be increased, either by increasing component reliability or by augmenting their redundancy. For generic scenarios, it will be enough to specify the unreliability of the systems and procedures involved.

5.   TRANSITION TO PRACTICE

The practice of assessing and controlling risk and detriment from potential nuclear accidents and abnormal radiological events is not, in itself, new. Some assessment techniques have been developed and safety is reflected in a number of codes and standards. Before approval to operate a facility is granted, a process of assessment for design, operation and maintenance of the facility is typically conducted from which the confidence is gained that established probability criteria are not likely to be exceeded. From this assessment it is possible to identify features of the facility which govern failure probabilities. Such features include performance and reliability of equipment and systems, technical specifications, requirements for testing and maintenance, training, etc.

The adequacy of a performance assessment seems to center on several key questions. Is there enough information to make to make a creditable estimate? Does the assessment cover a reasonable range of sensitivity and uncertainty analysis? Is the assessment complete? Does it show that performance criteria are likely to be met? Although performance assessments of nuclear power plants and other nuclear facilities have been conducted for some time, a number of concerns have been expressed about ICRP including potential exposure in its radiation safety recommendations. The central issue seems to be whether or not the questions just noted can be answered with the degree of confidence required in a regulatory regime which incorporates the ICRP recommendations. The three problem areas most often identified are assessment techniques, human reliability and demonstration of regulatory compliance.

5.1   Assessment Techniques

There is a need to have standard techniques available to assess risk and detriment for demonstrating compliance with requirements related to potential exposure. Such techniques have been established for normal exposure situations, e.g. 'standard man' parameters for internal dose assessments and standard dose weighting to calculate effective dose equivalent. Standard techniques can be established for assessments of potential exposure situations as well, and they could be used to determine the probability of situations leading to radiation exposure. Although the probabilistic techniques for calculation of safety system's reliabilities are currently well established, it appears that when taking into account factors such as common-mode failures and accident evolution, and after exercising explicit expert judgement, the results produced by different experts may vary widely even for the same scenario. For this reason, standard procedures for the quantitative assessments of potential exposure situations are needed in order to determine the probability that safety systems will fail to prevent a sequence of events having the potential to cause radiation harm. If such standard procedures are established, the inherent uncertainties which will result from their application must also be recognized.

5.2   Human Reliability

In some situations human performance will be the determining factor in the probability of a potential exposure occurring. Human reliability is difficult to model and quantify. Although dependence on human reliability to prevent potential exposures has in many cases been reduced, mainly by improved engineering, in cases where human performance dominates it will still be necessary to assign standard values for human reliability on the basis of the data available and by expert judgement.

5.3   Demonstration of Regulatory Compliance

Riks Limitation. In the case of normal exposure situations, the regulatory limits are expressed in terms of radiation doses and compliance with requirements can be demonstrated with adequate accuracy by the measurement of such doses or of dose-related quantities; doses can be recorded and regulatory conformance can be formally demonstrated (2). In the case of potential exposure situations, regulatory limits would likely be expressed in terms of probabilities and, since these cannot be measured directly, compliance for regulatory purposes cannot be demonstrated directly. There are, however, ways to provide reasonable assurance of compliance with requirements. Success depends to an important extent on how the requirements are framed and applied.

A simplified way to express limits for potential exposure could be accomplished by adopting a kind of criteria curve of probability versus dose. For instance, if the limit of probability of harm for the critical group is U_R, such criterion curve would have the shape as shown in Figure 1. The relevant features of the criterion curve are as follows: an inverse proportionality region; a non-proportional region for the dose range in which deterministic effects may also occur; and a constant probability for doses that are lethal.

In the lethal dose range, the probability is constant irrespective of dose, because the consequence to the invidual is the same regardless of the dose received. For the range of doses in which only stochastic effects occur, the relationship between probability and dose is inversely linear, with values representing the product of the probability of the dose, the annual dose, and the probability of a health effect per unit dose. Finally, in the dose range where deterministic effects may occur, i.e. individual doses exceeding a few sievert, the shape of the criterion curve is non-linear, in order to take into account the increasing probability of death. (This portion of the curve should approximate a sigmoid relationship and would depend to some extent on the time over which the dose is delivered).

The proposed criterion curve might be used to indicate whether a given safety option complies with the individual-related requirements in the following manner. First, the events, or sequences of events, with the potential to cause exposure to individuals should be identified. An event or sequence of events might be selected as representative of a group of similar scenarios, so long as the maximum consequences are considered. Second, the probability of occurrence of each event, and the consequent exposures of the critical group should be assessed, e.g. by taking into account the reliability of the safety systems and procedures. Finally, the point representing the probability of occurrence of the initial event and all other environmental conditions and the corresponding maximum dose is plotted. If the point is in the unacceptable region, then the option should be rejected. However, even if all the points are in the acceptable region, the proposal being assessed is not automatically acceptable because it may

---

(2)   As a practical matter, it is probably desirable to control risk attributable to normal exposure separately from risk attributable to potential exposure. The relevant dose limits and safety practices required to ensure compliance with the system of dose limitation for normal exposure situations need not be changed in a consolidated system for radiation safety.

be not optimized. Therefore, the usefulness of a criterion curve is limited at this stage to that of a basic decision tool for checking whether an option is unacceptable.

When a facility becomes operational, demonstration of compliance with reliability requirements presents a different kind of problem. As noted previously, unlike dose, reliability cannot be directly monitored. However, the operator can demonstrate compliance for regulatory purposes by demonstrating that the technical specifications and procedures are observed, that required maintenance is performed, and that safety equipment and systems function with acceptable reliability through appropriate tests and checks.

Optimization. The next stage is to check whether the option meets the ultimate requirement that safety is optimized. Regulatory compliance with the optimization requirement can be demonstrated by following a pre-specified, structured approach to optimization of safety. This will ensure that no important aspects are overlooked and to record the analysis for information and for assessment by others.,

The essential elements of this structured approach can be set out as shown in Figure 2. The terms are used as follows:

-   <u>Safety options</u> mean a specified design or a set of operational procedures;

-   <u>base case</u> means the starting point from which changes are assessed (in a design study, the base case normally is the cheapest option; in the case of operations, the base case is the current set of procedures);

-   <u>factor</u> means an identified measure of an option;

-   <u>safety factors</u> mean those factors which are related to the level of safety achieved (these will include those factors describing the resultant distribution in doses and their occurrence probability in any way, and those factors describing the safety efforts, such as costs and other disadvantages, incurred in modifying such distribution);

-   <u>other factors</u> mean those factors which are related to or describe the performance or costs of an option but are not related to the level of safety;

-   <u>performance</u> of <u>the option</u> means the results of applying a specified design option or set of operational procedures (expressed in terms such as the resultant doses and occurrence probabilities);

-   <u>criterion</u> means a quantitative or qualitative measure of what is acceptable or desirable for one or more of the factors.

6. CONCLUSION

The purpose of including potential exposure situations in the ICRP system of radiation safety is to provide a more comprehensive and coherent system of protection. In reality it is a recognition of what has been done in the past to some extent, although its inclusion in the ICRP recommendations may result in a more rigorous, systematic and balanced approach to overall safety. Its introduction, however, involves a number of complex issues and the transition to practice within the ICRP system will be evolutionary. Even after consensus is reached on methodologies to apply the ICRP recommendations in practice and as data bases improve with time, it is unlikely that assessments of potential exposure will stand as unequivocal proof that performance standards will be met and safety conditions optimized. The more likely result of such assessments will be a complex set of results which requires the exercise of careful judgement to discern acceptability within standards. This, however, should not stand in the way of attempting to improve our understanding of radiation risks and thereby enhancing the level of safety which can be provided to workers and the public.

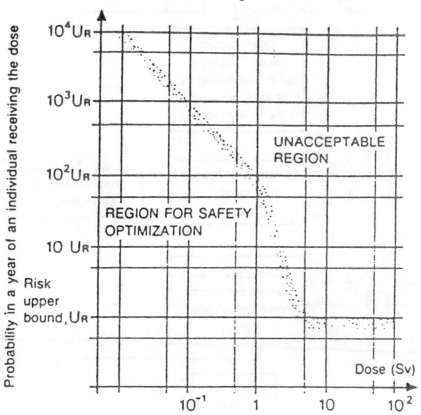

Figure 1

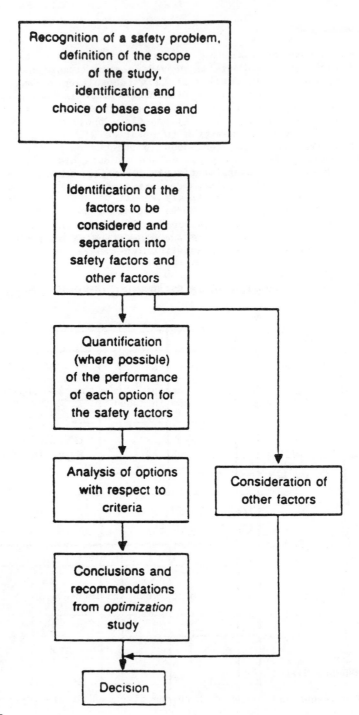

Figure 2

DISCUSSIONS

D. BENINSON, ICRP

The paper is now open for discussion.

A.P.U. VUORINEN, Finland

The logic of the terms normal and pre-existing exposure situations is not clear and I would like to get further clarification. Especially I am interested to know the reasons for separating these two cases. It is evident that protection approaches differ from each other on the practical point of view. Does the ICRP consider differently an exposure of 1 manSv in these two situations mentioned?

R.E. CUNNINGHAM, United States

The terms "normal", "pre-existing" and "potential" exposure situations were used in an earlier draft of the revised ICRP recommendations. These terms have now been changed to make the concept of how they are applied more clear. Normal and potential exposure are now included in the term, "practice". Adding a practice implies adding a risk. "Intervention" is currently being used instead of the term, "pre-existing". It implies that the risk exists in the environment and the decision centers on subtracting a risk by intervention. Decisions about adding a risk and subtracting a risk are fundamentally different.

D. BENINSON, ICRP

I agree. In one case you are planning to do something that would increase doses to people. They already receive doses, and, you must justify the introduction of the new practices. You should then do anything you can, reasonably, to reduce the doses from the practice. This is called optimisation.

In the other cases, you are not going to increase doses. The question is, are you concerned enough that you want to do something. In the case where you do an action, you decrease doses that already exist. It is different to cure something and to plan something.

J.F. CAMPBELL, United Kingdom

The issues and concepts which one discussed in the paper of Dr. Cunningham and Dr. Gonzalez are ones which are familiar to the nuclear safety community. Solutions, or rather working positions, have been hammered

139

out, generally by protacted negotiations between licensing authorities and licensees.

For example, in considering the possible application on accidents, the qualitative factors generally dominate the decision making process and hence such formal methods find little use in practice.

Looking through the ICRP proposals, I only see one point where they might impact nuclear safety, if adopted, and that is the question of adopting a common criterion on individual risk. In this connection, it is very important to distinguish between mandatory limits, which are referred to by radiological protection people, and targets or objectives (non-mandatory), with which the nuclear safety community is mainly concerned.

A. GONZALEZ, IAEA

The authors have recognised the lack of novelty of the different concepts and criteria presented in the paper (see point 5 of the report, first paragraph). They feel, however, that putting all of them together in a logical construction will enhance the coherence and consistency of the currently fractioned practices for controlling radiation risk. Thus, the "solutions ... hammered out ... by protracted negotiations between licensing authorities and licensees", as Mr. Campbell refers to, will be supported by common and logical guiding principles.

Moreover, multiattribute analysis is not in contradiction with the use of qualitative factors. Just the contrary. It allows the use of qualitative attributes in a coherent and traceable manner.

Finally, a distinction should be made not only between limits and targets or objectives, but also between basic limits (e.g. individual dose and risk limits) and derived limits which are the ones really applied in practice (e.g. limits of intake, for "normal" situations; limits on the unreliability of safety systems, for "potential" situations). Our paper refers to basic limits (rather than objectives or targets) but also indicates that in practice derived quantities will be used in practical implementation.

R. NAEGELIN, Switzerland

In your paper you use the term "Optimization" only and not the term "ALARA". Optimization is often understood as an analytical procedure similar to "cost benefit" calculations. As regulators we prefer the "ALARA" principle, putting, the risk in the center of consideration, that should be as low as practically achievable, while in "optimization" cost considerations could have more weight.

R.E. CUNNINGHAM, United States

We understand optimisation to mean the same thing as ALARA, as does the ICRP.

UN POINT DE VUE SURETE NUCLEAIRE
SUR L'APPROCHE CIPR POUR LES EXPOSITIONS POTENTIELLES
ET SUR LE ROLE DES ANALYSES PROBABILISTES DANS CE DOMAINE

Pierre Y. Tanguy

Résumé

Dans le projet de recommandations diffusé par la CIPR au début de l'année 1990, l'extension du système de protection aux expositions potentielles constitue une novation importante. Des réserves ont été faites par le Comité Consultatif International pour la Sûreté Nucléaire de l'AIEA et seront reprises, avec des commentaires personnels de l'auteur.

Les possibilités offertes par des Evaluations Probabilistes de Sûreté pour des centrales nuclaires seront passées en revue. Leurs limites et leur rôle pour la sûreté seront discutées en fonction du niveau de l'évaluation.

Au début de 1990, la Commission Internationale de Protection Radiologique (CIPR) a largement diffusé dans la communauté scientifique un projet de recommandations[1]. Le Comité Consultatif International sur la Sûreté Nucléaire, connu sous le sigle INSAG, que l'AIEA a établi pour conseiller le Directeur Général, a reçu copie de ce projet, et a transmis le 2 avril 1990 son avis au Président de la CIPR, D. Beninson, également membre de l'INSAG.

Ma présentation reprendra, avec des commentaires personnels, certains passages de cet avis. Comme le sujet traité, expositions "potentielles" aux rayonnements ionisants, implique naturellement le concept de probabilité, cette analyse du projet CIPR sera complétée par une discussion des possibilités, et des limitations, des études probabilistes de sûreté dans le domaine de la protection des travailleurs, du public et de l'environnement vis-à-vis des risques liés à des accidents nucléaires.

I - LE PROJET CIPR

Une innovation importante du nouveau projet par rapport aux recommandations antérieures de la CIPR, et vis-à-vis des règles gouvernant habituellement les relations entre la sûreté nucléaire et la radioprotection, est l'extension du cadre général de la protection radiologique à ce que la CIPR appelle les "potential exposure situations". Les dispositions proposées portent alors sur l'importance des expositions, c'est à dire sur les doses, mais aussi sur la probabilité d'occurrence des situations. Cette notion est introduite au §108[2], et le §110 indique explicitement que les trois principes de base, justification, optimisation et limitation, s'appliquent à ces expositions potentielles.

Dans un document transmis en mars au Comité Scientifique et Technique d'Euratom[3], H.J. Dunster reconnait la novation que représente cette introduction des expositions potentielles par rapport aux errements antérieurs. Il écrit que "ICRP is attempting

---

[1] Depuis la diffusion de ce document, la CIPR s'est réunie à Washington en juin. Le communiqué de presse diffusé le 25 juin mentionne les "practices giving rise to radiation exposures", mais ne parle pas explicitement des expositions potentielles.

[2] Je me réfèrerai à la numérotation du document en date du 25 janvier 1990.

[3] STC-90-D-126

to provide a link (between nuclear safety and radiation protection) by entering the field of prevention of accidental exposures...".

Cette extension du système de protection CIPR est développée dans les §4.3, 4.6 et 5.6. Le §4.3 traite rigoureusement sur le même pied les expositions programmées - "planned exposures"- et les expositions potentielles. Le §4.6 est supposé donner des précisions particulières aux expositions potentielles, mais distingue en fait essentiellement trois cas: exposition suffisamment probable pour qu'on puisse l'identifier à une situation programmée; situation moins probable, mais qui s'est finalement produite, et qui est traitée comme une "situation pré-existante"; et situation qui n'a qu'une probabilité limitée de se produire. Le §5.6 traite spécifiquement le cas de ces dernières. C'est lui qui va nous intéresser avant tout du point de vue sûreté, sous les deux angles: prévention et mitigation.

Le texte de la CIPR (§190) distingue effectivement les deux aspects, prévention et mitigation, mais dans un sens un peu différent de celui qui est généralement utilisé en sûreté, ce qui peut prêter à confusion. C'est ainsi que l'événement considéré étant l'exposition aux rayonnements, les structures de confinement d'une installation nucléaire, qui préviennent la fuite de la radioactivité vers des individus, relèveraient de la prévention, plutôt que de la mitigation. Il en serait de même d'une évacuation préventive des populations. Le point essentiel c'est que, reconnaissant qu'il faut traiter séparément probabilité d'occurrence et magnitude de la dose ("a reduction in probability by a factor is not equivalent to a reduction in dose by the same factor"), la CIPR considère que le seul critère doit porter sur le risque, qui combine deux probabilités: pour un individu, celle de recevoir une dose de radiation, et celle que cette dose entraine une mort par cancer. Cette combinaison regroupe donc ce que la sûreté traite séparément en prévention et mitigation, et élimine ainsi automatiquement un des principes de base de la sûreté, la défense en profondeur: faire tout pour éviter l'accident, mais supposer qu'il se produit quand même et s'attacher à ce que ses conséquences restent limitées.

La CIPR introduit alors la notion de "contrainte" de risque, en dessous de la limite proprement dite, qui est une facilité de gestion des projets, qui a des équivalents plus détaillés dans les objectifs fiabilistes de la sûreté; j'y reviendrai plus loin.

La difficulté majeure, que reconnait d'ailleurs tout-à-fait le projet de la CIPR, apparaît quand on veut dépasser le risque individuel et passer au risque collectif. La spécification pour le détriment collectif, mesure des conséquences potentielles, est difficile et controversée (§193). La CIPR indique que le risque, défini comme le produit probabilité-conséquence, implique justement cette réciprocité entre les deux termes qu'elle a dénoncée. Elle semble alors recommander l'analyse multi-critères ("multi-attribute analysis") avec ou sans ce que j'appellerai une "concaténation" ("aggregation of weighted attributes"). Il faut noter que l'emploi du détriment collectif est demandé dès l'application du premier principe, la justification (§196).

La limite proposée pour le risque individuel (§198) ne semble pas recouper les objectifs de sûreté discutés au plan international: elle correspond pour le public à une probabilité de décès de 5.10^{-5} par an. Mais il faudrait examiner de plus près ses modalités d'application pour en comprendre la signification pratique.

II - L'AVIS INSAG

L'INSAG a exprimé de sérieuses réserves sur ce qui est proposé en matière d'expositions potentielles et a finalement pris clairement position contre cette extension de la compétence de la CIPR à la sûreté nucléaire.

Il lui a semblé en effet que l'application du système de protection CIPR, et en particulier de ses trois principes fondamentaux, justification, optimisation et limitation, ne se justifie que pour les situations à relativement forte probabilité, celles par exemple que l'on doit s'attendre à voir se produire pendant la durée de vie d'une installation nucléaire (probabilité $> 10^{-2}$ pour fixer les idées). Dans ce cas, l'INSAG est d'accord avec le projet (§130), qui prévoit que de telles situations doivent être traitées comme si elles étaient certaines.

On doit par contre considérer que l'application du principe d'optimisation à des situations de faible probabilité (accidents de fréquence faible et accidents graves hypothétiques dans le langage de la sûreté) n'est pas acceptable. Les incertitudes inhérentes à l'analyse de telles situations, tant sur leur probabilité que sur leurs conséquences, peuvent en effet faire craindre qu'une approche par trop théorique ne conduise à des conclusions qui aillent à l'encontre de la recherche du meilleur niveau de sûreté. L'absence de "coupure", tant en probabilité qu'en conséquence, peut conduire à des évaluations sans signification réelle, à partir du moment où elles se ramènent à des produits de zéro par l'infini, avec des incertitudes toujours croissantes sur les deux termes.

Le projet CIPR reconnait (§193&194) que la notion de détriment collectif est très controversée. La suggestion d'utiliser des méthodes muti-critères ne fait que déplacer le problème. Le document CIPR ne fait d'ailleurs pas progresser la question de l'évaluation du détriment collectif dans le cas où de très faibles doses s'appliqueraient à un grand nombre d'individus, ce qui est précisément une éventualité à envisager dans les expositions potentielles de faible probabilité, que ce soit pour le public éloigné, géographiquement, en cas d'accident sur une installation nucléaire, ou pour les populations lointaines, dans le temps, pour les effets liés à des stockages de déchets radioactifs.

On espérait une position constructive sur la question capitale de "l'exemption du contrôle règlementaire", qui rejoint ce qui a été souvent appelé les niveaux "de minimis". Cette question est traitée au §7.7. La CIPR refuse de prendre une responsabilité dans ce domaine, alors que la communauté nucléaire a besoin de valeurs acceptées internationalement. Le §277 est particulièrement décevant de ce point de vue. On peut se demander pourquoi la CIPR n'a pas cru devoir au moins donner son aval aux niveaux 10 microSievert pour la dose individuelle, et 1 homme x Sievert pour la dose collective, retenus par l'AIEA. Il est clair que la CIPR se sent liée par son affirmation catégorique, et répétée, de toute absence de seuil (§66). Elle refuse la notion de seuil "pratique", qui correspondrait en particulier au point où la probabilité d'induction de cancer devient si faible qu'elle perd toute signification individuelle. Une telle position serait analogue à celle prise par les experts de sûreté lorsqu'ils considèrent qu'en dessous d'un certain niveau de probabilité d'occurrence, un évènement n'a plus à être pris en compte dans l'analyse de sûreté.

Cela me parait personnellement une des faiblesses du projet de recommandations, d'autant que l'on sait en outre que c'est précisément cette absence de tout niveau en dessous duquel il n'y a "en pratique" aucun risque sanitaire réel qui empêche le public d'avoir une juste vue du risque radioactif.

Au cours d'une réunion de travail au sein de l'INSAG, le Président D. Beninson a déclaré que pour chaque type de situation il y a un des trois principes qui doit s'appliquer de manière dominante:

- pour les expositions certaines (planned exposures), c'est l'optimisation, que tout le monde désigne sous le nom ALARA (les expositions restant bien sûr toujours en dessous des limites);
- pour les expositions pré-existantes, c'est la justification des pratiques (faut-il ou non faire quelque chose?);
- pour les expositions potentielles, c'est la limitation, limitation du risque individuel en terme de probabilité de décès (Cf. §198).

Cette conception a sa logique, mais elle ne figure pas dans le projet CIPR actuel. De plus, en ce qui concerne les situations de faible probabilité, on ne voit pas ce que l'application du système CIPR apporte comme valeur ajoutée par rapport aux principes de sûreté nucléaire qui font l'objet d'un consensus international. Le document INSAG-3, Principes Fondamentaux de Sûreté Nucléaire, publié en 1988, inclut des données qui recouvrent les deux principes que la CIPR appelle justification et limitation. Leur extension risquerait donc de créer une confusion dans les esprits, préjudiciable à une bonne sûreté. Je reviendrai plus loin sur la question des objectifs de sûreté, qui constituent le pendant, côté sûreté nucléaire, des contraintes ("constraints") et limites du système CIPR.

L'INSAG a souhaité enfin une clarification du vocabulaire. L'utilisation du mot "inacceptable" qui apparait pour la première fois au §121 donnera lieu dans le public à toutes les erreurs d'interprétation mentionnées d'ailleurs dans le projet (§122). Le terme est repris au §147 et l'introduction du concept de "tolérabilité", §148, n'arrange pas les choses, mais peut au contraire ajouter à la confusion. Certains termes utilisés, §158 à 160, "just short of unacceptable", "probably too high and regarded by many as being so", montrent à l'évidence qu'il ne s'agit pas d'une inacceptabilité en terme sanitaire seulement.

III - REMARQUES COMPLEMENTAIRES

Le projet CIPR traite aussi de la question de l'interférence éventuelle entre situations certaines et potentielles, évoquée aux §144 et 197, d'ailleurs de façon pas très cohérente, et traitée au §201. Je considère personnellement que la CIPR devrait renoncer à faire de recommandation sur un sujet particulièrement complexe et qui a jusqu'ici été traité au cas par cas sans avoir besoin de théorie générale.

J'ai eu connaissance de trois documents qui analysent la question des expositions potentielles telle que la présente le projet CIPR.

Le premier[1]émane du Commissariat à l'Energie Atomique et regroupe les opinions de scientifiques et ingénieurs français impliqués dans les activités nucléaires nationales. Avec des mots différents, il reprend sensiblement les mêmes idées que l'INSAG.

Le deuxième[2] vient du NRPB. Il indique que l'introduction de limites et de contraintes de risque est une addition bienvenue, mais reprette l'absence de justification des valeurs proposées, qui, selon lui, devraient résulter d'une discussion sur la "tolérabilité" des risques.

Le troisième m'a été transmis par la Division de la Sécurité des Installations Nucléaires en Suisse[3]. Il appelle l'attention sur le fait que la conformité avec le principe d'optimisation doit faire appel aux techniques d'études probabilistes de sûreté (Probabilistic Safety Assessment, PSA) et en discute les possibilités et les limites. J'en reprendrai quelques éléments dans la suite de cette présentation.

IV - LES ETUDES PROBABILISTES DE SURETE

Je me limiterai au cas des seuls réacteurs électro-nucléaires. Le rapport INSAG-3 déjà cité fait une large place aux études probabilistes et permet de situer à quels niveaux elles se situent dans l'évaluation du risque et l'analyse des dispositions de sûreté.

L'objectif général de sûreté nucléaire étant de protéger les individus, la société et l'environnement contre le risque radiologique, le rapport définit l'objectif de sûreté technique: prévenir les accidents et faire en sorte que la probabilité d'accidents graves avec conséquences radiologiques importantes soit extrêmement faible. Le commentaire fixe le but à atteindre: ne pas dépasser 10^{-5} pour la probabilité d'endommagement grave du coeur, et réduire d'un facteur 10 au moins la probabilité de rejets nécessitant une intervention rapide hors site, grâce aux mesures de mitigation. On voit ainsi apparaître les deux premiers niveaux des études probabilistes de sûreté:

- niveau 1: évaluation du risque d'endommagement de coeur, assimilé souvent à la fusion;

---

[1] Référence HBR/BBD-H311 du 26 mars 1990

[2] Référence M-238

[3] Y.G. Gonen - HSK

- niveau 2: évaluation du risque de rejets radioactifs à l'extérieur du site, qui donne par rapport au niveau 1 la probabilité conditionnelle de rejets extérieurs associée à l'importance des rejets.

Le rapport INSAG-3 n'introduit pas d'objectif de sûreté lié au risque individuel ou collectif, qui impliquerait une étude de niveau 3, c'est à dire une évaluation des conséquences sanitaires et économiques des rejets sur le public et dans l'environnement.

L'analyse probabiliste est également citée dans le rapport INSAG-3: dans les principes fondamentaux, en soulignant sa complémentarité avec l'analyse de sûreté déterministe tant pour définir les dipositions de sûreté que pour en évaluer les mérites; dans les principes particuliers, par exemple à travers les objectifs de fiabilité des équipements et des systèmes, et les agressions externes à prendre en compte dans la conception des centrales.

Ce très bref survol me permet de souligner deux points importants:
- les études probabilistes de sûreté d'une installation donnée, telles celles qui ont été effectuées dans de nombreux pays[1], qui permettent d'obtenir une évaluation globale de risque, et auxquelles la fin de cette présentation sera consacrée, ne constituent qu'une partie des études probabilistes utilisées dans la sûreté nucléaire;
- la sûreté nucléaire ne repose pas uniquement sur des analyses probabilistes, mais fait toujours largement appel aux méthodes déterministes, qui permettent en particulier de protéger une installation contre les risques techniques identifiés en bénéficiant de l'expérience industrielle acquise.

Dans une discussion sur les possibilités et les limites des études probabilistes de sûreté relatives à une centrale, que je dénommerai désormais par leur sigle anglais PSA, il me parait nécessaire de distinguer les niveaux.

NIVEAU 1

Au niveau 1, risque de fusion de coeur, je pense qu'il existe aujourd'hui dans le monde de nombreuses études qui ont fait l'objet de revues critiques détaillées et qui ont acquis une grande

---

[1] Depuis le rapport Rasmussen de 1975 qui fait toujours référence, je citerai Biblis en RFA, 5 centrales US comparées dans le rapport NUREG-1150, la Suède, la Suisse, etc. et tout récemment en France, la publication en mai 1990 de deux PSA niveau 1, sur un PWR 900 et sur un PWR 1300.

crédibilité. Je crois qu'il faut souligner que ce sont des études "lourdes", qui ont exigé une analyse très détaillée non seulement de la conception de la centrale, mais aussi de son fonctionnement à travers toutes les procédures d'exploitation, et qui se sont appuyées sur des données de fiabilité, tant sur les matériels que sur les hommes, ayant fait l'objet d'examens critiques pour s'assurer qu'elles étaient bien représentatives de la centrale étudiée. Ces PSA sont aujourd'hui généralement "transparents", c'est à dire qu'on peut assez aisément procéder à des intercomparaisons, au niveau des hypothèses de base, des modèles de calcul et des résultats. Ces derniers restent entachés d'une marge d'incertitudes, mais qui est assez réduite aujourd'hui; je pense que dans un PSA de qualité, pour la probabilité globale, et en se limitant aux défaillances et agressions internes à la centrale, l'intervalle 5%-95% est de l'ordre d'une décade. Je pense qu'on peut donc affirmer que nous disposons aujourd'hui d'un outil au point et utilisable aussi bien pour des études particulières de sûreté, j'y reviendrai plus loin, que pour des évalautions globales de risques.

J'ai fait une réserve en ce qui concerne les agressions externes. Je pense qu'il faut attendre les conclusions de la "Peer Review" en cours du rapport NUREG-1150, conduite par des experts internationaux sous la direction de H. Kouts, pour savoir si on peut être aussi affirmatif que je l'ai été dans le cas de l'évaluation "interne", en particulier en ce qui concerne la prise en compte des séismes.

En ce qui concerne les autres limites à associer aux résultats des PSA niveau 1, je considère pour ma part que sous réserve que l'équipe chargée de l'étude ait une bonne connaissance du fonctionnement normal et accidentel de l'installation, les progrès faits ces dernières années ont considérablement réduit les incertitudes citées généralement et que je rappelerai ci-dessous:
- la non-exhaustivité des événements initiateurs: l'expérience accumulée nous donne une garantie d'exhaustivité me semble-t-il pour les événements relativement fréquents, $>10^{-3}$ en ordre de grandeur, et la prise en compte déterministe des différents modes de défaillances rares rend peu probable qu'un type nouveau puisse conduire à une séquence dominante.
- la prise en compte des modes communs: dès lors que les analystes ont approfondi les interactions possibles entre systèmes, il existe aujourd'hui des méthodes, appuyées sur l'expérience, qui donnent à mon avis d'assez bons résultats; c'est néanmoins un des points où une attention particulière à tous les événements

d'exploitation est indispensable pour identifier d'éventuels modes communs oubliés. Il en est de même pour les scénarios d'accident: l'expérience peut révéler des embranchements insoupçonnés.

- la prise en compte du facteur humain: notre expérience à EDF nous a montré qu'à partir des résultats d'essais sur simulateurs, il était possible d'arriver à une bonne connaissance tant qualitative que quantitative des comportements humains, en opération et en maintenance, qui porraient initier ou aggraver des séquences accidentelles. Il faut cependant souligner que cela demande un travail important.

Niveau 2

Alors que les études accidentelles s'arrêtent dans les PSA niveau 1 au moment où le cœur n'est plus refroidi convenablement, le PSA niveau 2 impose que soit poursuivie l'étude du déroulement de l'accident dans ses phases utérieures jusqu'à l'atteinte d'une situation définitivement stabilisée. Ensuite, il faut évaluer le comportement de l'ensemble des structures de l'installation qui vont jouer un rôle dans le confinement des produits radioactifs, et en particulier celui de l'enceinte de confinement avec tous ses "by-pass" possibles, dans des conditions pour lesquelles ces structures n'ont pas été prévues. Enfin, il faut suivre le cheminement des produits radioactifs eux-mêmes, qui sera fonction pour les différents radionucléides de leur état physico-chimique et de leur évolution. Il est bien clair que ceci modifie du tout au tout les conclusions auxquelles j'étais arrivé pour le niveau 1:

- les incertitudes deviennent beaucoup plus considérables, et sont de plus difficiles à quantifier dans la mesure où pour beaucoup des problèmes qui se posent et que je vais énumérer ci-après, on est très dépendant de jugements d'experts;

- les modèles décrivant les phénomènes physiques sont souvent entachés de sérieuses insuffisances; les scénarios ne peuvent plus toujours être établis avec les embranchements par oui ou par non que l'on utilise au niveau 1; dans certains cas, ils doivent être regroupés par types de phénomènes et perdent leur identité particulière;

- on manque de données non seulement pour quantifier certaines séquences, mais parfois pour les définir avec suffisamment de détails; il est donc difficile de garantir l'exhaustivité;

- on est enfin obligé de poursuivre l'étude sur de longues durées, ce qui l'alourdit, sans que l'on ait l'assurance de prendre correctement en compte les différentes actions de gestion d e l'accident à la disposition des opérateurs.

Je pense que pour tous ces aspects, le rapport américain NUREG-1150 et la critique qui en sera faite constitueront pour plusieurs années la référence internationale. Il ne me semble pas qu'il faille espérer des résultats beaucoup plus convaincants d'autres études de même type. Par contre, j'estime personnellement qu'il y a beaucoup à apprendre d'études probabilistes partielles, dans lesquelles on s'efforcerait d'analyser de manière approfondie le comportement du confinement pour quelques séquences de référence qui paraissent a priori les moins hypothétiques. Renonçant ainsi délibérément à l'exhaustivité, il serait néanmoins possible d'évaluer certains points faibles et d'y remédier.

Niveau 3

Il s'agit d'évaluer les conséquences finales. On retrouve naturellement au départ toutes les incertitudes du niveau 2, tant en ce qui concerne les probabilités que les termes-sources. Il s'y ajoute les problèmes liés au transfert des radionucléides dans l'environnement, à l'efficacité des plans d'urgence et de réhabilitation, et aux effets des faibles doses.

Les résultats du rapport NUREG-1150, exprimés en risque annuel de décès précoce et de cancer différé, comparés à la référence Rasmussen, ne me paraissent personnellement pas apporter d'enseignement vraiment significatif pour la sûreté. Pour le public, je ne vois pas comment on peut tirer parti de ce que pour Grand Gulf et Peach Bottom par exemple on gagne environ trois décades sur le risque de décès précoce évalué en 1975, mais seulement une décade sur le risque de cancer!

Contrairement à ce que j'ai indiqué pour le niveau 2, je ne vois pour ma part guère d'intérêt à poursuivre les PSA niveau 3, même de manière partielle. Je rappelle que l'application des recommandations de la CIPR concernant les expositions potentielles impliquerait précisément de pousser systématiquement à ce niveau.

V - LE ROLE DES PSA

Je classerai en 5 rubriques l'utilisation possible d'un PSA:
- pour la conception d'un nouveau réacteur; à mon avis, comme on ne peut disposer au départ d'une description détaillée de l'installation et de son mode d'exploitation, il ne peut s'agir que d'un PSA "en différentiel" par rapport à un réacteur en service; si le type de réacteur est carrément révolutionnaire, et ne peut se rattacher à

un modèle existant, je ne pense pas qu'il faille parler de PSA à proprement parler, mais d'évaluation partielle;

- pour l'exploitation d'un réacteur en service, en identifiant les séquences dominantes, comme aide à la décision pour les modifications à apporter à l'installation ou à ses procédures, de conduite, d'entretien et d'essais périodiques;

- dans le dialogue entre exploitant et autorité de sûreté pour toutes les questions de sa compétence; le PSA ne doit pas être le "juge de paix", mais il apporte une clarification utile au débat technique;

- dans l'élaboration des programmes de recherches en aidant à identifier les points où une amélioration des connaissances est utile ou nécessaire;

- éventuellement pour mettre de la rigueur dans le débat démocratique, mais en l'utilisant avec une grande modestie.

Mais pour qu'un PSA puisse jouer pleinement son rôle, il doit à mon avis remplir plusieurs conditions:

- être au départ indiscutable sur le plan de la qualité et de la transparence; méfions-nous des PSA "légers", n'ayant pas fait l'objet d'une revue critique, et dont les rouages ne sont pas accessibles à des experts extérieurs;

- être tenu à jour en permanence, tant au niveau des données grâce à un accès à des banques de données de fiabilité, matérielle et humaine, qu'à celui des modèles et des scénarios, en faisant notamment usage du retour d'expérience; ceci implique que sous l'angle informatique en particulier, le PSA soit prévu pour faciliter cette tenue à jour;

- être mis sous forme conviviale pour pouvoir être utilisé par des techniciens, formés spécialement certes, mais qui n'aient pas à être des spécialistes de fiabilité et de probabilités.

Les PSA jouent déjà aujourd'hui un rôle essentiel pour la sûreté. Je pense que d'ici quelques années leur emploi se généralisera, même si les bonnes vieilles méthodes déterministes et l'application du principe de défense en profondeur garderont toute leur importance.

A NUCLEAR SAFETY VIEW ON THE PROPOSED ICRP APPROACH TO PROBABILISTIC EXPOSURES AND THE POSSIBILITIES AND LIMITATIONS OF USING PRA IN THIS FIELD

Pierre Y. Tanguy
Inspecteur Général, Electricité de France
Paris, France

ABSTRACT

The extension of the ICRP system of radiation protection to encompass potential exposures in the new recommendations published by that body in early 1990 represents a significant innovation. The author reviews and provides a personal commentary on the reservations which the International Consultative Committee for Nuclear Safety of the IAEA has voiced with regard to these recommendations.

The paper also reviews the potential contribution that Probabilistic Safety Analyses can make in nuclear power plant safety management, and discusses their limitations and role according to level of assessment.

In early 1990, the International Commission on Radiological Protection (ICRP) widely circulated copies of new draft recommendations among members of the scientific community (1). The International Nuclear Safety Advisory Group (INSAG), set up to advise the Director General of the IAEA, was sent a copy of these recommendations and on 2 April 1990 submitted an opinion to the Chairman of ICRP, D. Beninson, who is also a member of INSAG.

My paper will review and comment on certain passages contained in the written opinion given by INSAG. Given that the subject matter, "potential" exposures to ionising radiation, must by definition touch upon the concept of probability, my analysis of the draft ICRP recommendations will conclude with a discussion of the potential applications and limitations of probabilistic safety studies with regard to the protection of workers, the general public and the environment against the risks associated with nuclear accidents.

I. THE NEW ICRP RECOMMENDATIONS

One significant innovation in the new recommendations, compared with previous ICRP recommendations and also the rules that normally apply to the relationship between nuclear safety and radiation protection, consists in extension of the general framework of radiological protection to encompass what the ICRP refers to as "potential exposure situations". The proposed recommendations therefore cover not only the size of exposures, i.e. dose rates, but also the probability of such situations occurring. This concept is first introduced in §108 (2), and §110 explicitly states that the three basic principles of justification, optimisation and limitation are applicable to such potential exposures.

In a document submitted in March to the Scientific and Technical Committee of Euratom (3), H.J. Dunster recognised that introduction of the concept of potential exposure situations represented a radical departure from previous practice. He commented that "ICRP is attempting to provide a link (between nuclear safety and radiation protection) by entering the field of prevention of accidental exposures ...".

The extension of the ICRP system of radiation protection is described in paragraphs 4.3, 4.6 and 5.6. Paragraph 4.3 specifically places "planned exposures" and "potential exposures" on an equal footing. Furthermore, while paragraph 4.6 is supposed to provide details regarding potential exposures, what it basically does is distinguish between three specific cases: exposures of sufficiently high probability to be identified with a planned situation; less probable situations which have ultimately occurred and which are dealt with as "pre-existing situations"; and situations whose occurrence is merely of limited probability. Paragraph 5.6 deals specifically with the latter. This

---

(1) Following publication of this document, ICRP held a meeting in Washington in June. The press release given out on 25 June refers to "practices giving rise to radiation exposures", but does not explicitly mention potential exposures.

(2) The numbering used here is that of the document dated 25 January 1990.

(3) STC-90-D-126.

is the paragraph that is primarily of interest to us in that it addresses two aspects of safety management: prevention and mitigation.

The ICRP recommendations (§190) do indeed make a distinction between these two aspects, prevention and mitigation, but in a sense that differs slightly to that in which it is used in the field of nuclear safety – which may lead to confusion. Since the recommendations consider events in terms of exposure to radiation, the provision of containment structures, which are designed to prevent radioactivity from reaching individuals, in a nuclear installation would therefore be a preventive rather than a mitigating measure. The same would be true of preventive evacuation of the population. The basic point here is that, recognising the need to deal separately with the probability of occurrence and magnitude of the dose ("a reduction in probability by a factor is not equivalent to a reduction in dose by the same factor"), ICRP considers that the only criterion should be that of risk, which for an individual consists of two probabilities: first, the probability of receiving a dose of radiation; second, the probability that such a dose will result in death from cancer. In combining two elements which in nuclear safety analysis are dealt with separately under the headings of prevention and mitigation, ICRP has automatically eliminated one of the basic tenets of safety – defence in depth: do everything possible to avoid an accident, but assume that it will occur just the same and try to ensure that it will be of limited consequence.

ICRP then introduces the concept of risk "constraint" below the risk limits as such, which is a way in which to facilitate project management and which has a more detailed counterpart in the reliability objectives of safety management. I shall return to this point later in the paper.

The main difficulty, as is clearly acknowledged in the new ICRP recommendations, is to be found in the transition from individual risk to collective risk. The specification regarding collective damage, the measure of potential consequences, is both complicated and controversial (§193). ICRP indicates that the concept of risk, defined as the product of probability and consequence, does indeed involve the very type of reciprocity between the two terms that it had claimed. It then seems to advocate the use of multi-attribute analysis, with or without what I would call "sequencing" ("aggregation of weighted attributes"). It should be noted that the recommendations call for collective damage to be considered as soon as the first principle, that of justification, is applied (§196).

The limit proposed for individual risk (§198) does not seem totally with the safety objectives discussed at international level; it corresponds, for the general public, to a probability of death of $5.10-5$ per year. We need to take a closer look at the way in which it is applied, however, if we are to understand what this means in practice.

II. THE OPINION GIVEN BY INSAG

INSAG voiced serious reservations concerning the proposals for potential exposures, and finally made it clear that it was opposed to extending the competence of ICRP to the nuclear safety field.

INSAG felt that application of the ICRP protection system, and in particular the three basic principles of justification, optimisation and limitation, could only be justified for situations which have a relatively high level of probability such as those that might be expected to occur during the lifetime of a nuclear installation (probability $> 10-2$, for example). On this point, INSAG is in agreement with the provisions of §130 of the draft which states that such situations should be treated as though they were certain to occur.

On the other hand, it would seem that applying the principle of optimisation to situations of low probability (infrequent accidents and hypothetical serious accidents, to use the jargon of safety analysts) is not acceptable. The uncertainties inherent in the analysis of such situations, in terms of both their probability and their consequences, must surely give rise to fears that too theoretical an approach might result in analysts drawing conclusions which would run counter to efforts to secure the highest level of safety. The lack of a "break", in terms of both probability and consequence, might produce assessments that, once they result in factors of zero to the power of infinity with increasing degrees of uncertainty on both terms, are of no real significance.

The ICRP draft recommendations (§193 and §194) acknowledge that the concept of collective damage is highly controversial. The suggested use of multi-attribute methods merely shifts the problem to another arena. Furthermore, the ICRP draft has made no progress on the assessment of collective damage in cases where large numbers of individuals are exposed to very low doses, which is precisely one possibility that should be considered with regard to potential exposures of either the general public in geographically remote locations, in the case of an accident at a nuclear installation, or future populations in the case of sites used for the storage of radioactive wastes.

We had hoped to see ICRP adopt a constructive stance on the crucial issue of "exemption from regulatory control", closer to what has often been termed "minimum levels". This issue is dealt with in paragraph 7.7. ICRP refuses to accept any responsibility in this domain, despite the fact that the nuclear community has a need for internationally accepted values. Paragraph 277 is particularly disappointing in this respect. We might well wonder why ICRP felt that it was under no obligation to endorse at least the 10 microsievert level for individual doses and the 1 man x Sievert level for collective doses set by the IAEA. ICRP clearly feels bound by its categoric and repeated declaration that there should be no threshold (§66). It refuses to accept the notion of a "practical" level, one corresponding in particular to the point at which the probability of inducing cancer is so low that it has no individual significance. Such a stance would be comparable to that taken by safety experts when they consider that a given event no longer needs to be taken into account in a safety analysis once it falls below a given level of probability of occurrence.

In my opinion, this is one of the weaknesses of the new recommendations; particularly when we consider that it is precisely this lack of a level below which there is "in practice" no real health risk that prevents the general public from gaining a clear understanding of the risks associated with radioactivity.

In the course of a working meeting at INSAG, the Chairman, D. Beninson, stated that for each type of situation there were three principles whose application was paramount:

- For planned exposures, the principle of optimisation; a principle that is commonly referred to as ALARA (exposures clearly remain below set limits at all times);

- For pre-existing exposures, the principle of justifying practices (should action be taken or not?);

- For potential exposures, the principle of limitation; limitation of individual risk in terms of the probability of fatalities (see §198).

This approach does have a certain logic, but it is not embodied in the current ICRP draft recommendations. Moreover, with regard to situations of low probability, it is by no means clear what kind of value added the ICRP system offers with regard to principles of nuclear safety on which an international consensus has been reached. The INSAG-3 report, Basic Safety Principles, published in 1988, contains data that cover the two principles referred to by ICRP as justification and limitation. Broadening the scope of these principles might simply lead to confusion and could have an adverse impact on safety. I shall return later in this paper to the question of safety objectives which are the counterpart, in terms of nuclear safety, to the constraints and limitations of the ICRP system.

Lastly, INSAG suggested that the terminology used by ICRP needed to be clarified. Use of the word "unacceptable", which first appears in §121, will encourage the public to make all the errors in interpretation referred to elsewhere in the recommendations (§122). The term is used again in §147, and introduction of the concept of "tolerability" in §148 does not help matters much and may well cause further confusion. Certain terms used in paragraphs 158 to 160 such as "just short of unacceptable" and "probably too high and regarded by many as being so" clearly indicate that what is deemed to be unacceptable is not restricted to health matters.

III. COMPLEMENTARY COMMENTS

The new ICRP recommendations also deal with the problem of interference between planned situations and potential situations, a problem that is mentioned, albeit in a somewhat garbled manner, in paragraphs 144 and 197, and discussed in greater detail in §201. My personal feeling is that ICRP should refrain from making recommendations about a particularly complex subject that until now has been dealt with on a case by case basis without any need for a general theory.

I have consulted three documents analysing the issue of potential exposures as presented in the ICRP draft.

The first (1) was published by the French Atomic Energy Commission and contains the opinions of French scientists and engineers involved in nuclear activities in France. Using different language, it basically expresses the same ideas as INSAG.

The second (2) was issued by NRPB. While welcoming the introduction of limits and constraints for risks, it regrets the lack of justification for the values proposed which, in its opinion, should have been decided through a discussion of the concept of risk "tolerability".

The third was sent to me by the Division de la Sécurité des Installations Nucléaires in Switzerland (3). It draws attention to the fact that compliance with the principle of optimisation must be based on application of the techniques used for probabilistic safety assessements (PSAs), whose potential and limitations it goes on to discuss. I shall make use of some of the points raised in this document later in my paper.

IV. PROBABILISTIC SAFETY ANALYSES

I shall limit my discussion to nuclear power reactors. The INSAG-3 report mentioned above devotes much of its discussion to probabilistic analyses and indicates the role they play in risk assessment and the analysis of safety rules.

Since the general objective of nuclear safety management is to protect individuals, society and the environment against radiological hazards, the report provides a definition of the objective of technical safety which is to prevent accidents and ensure that the probability of serious accidents with significant radiological consequences remains extremely low. The commentary identifies the goal to be achieved, which is to keep the probability of serious damage to the core at a level below 10^{-5} and, through the implementation of mitigating measures, to reduce by a factor of at least 10 the probability of a release requiring rapid action to be taken off-site. The first two levels of probabilistic safety analyses are therefore:

- Level 1: assessment of the risk of damage to the core, often equated with core melt;

- Level 2: assessment of the risk of radioactive releases off-site, which, with regard to level 1, provides the conditional probability of external releases associated with the size of the releases.

The INSAG-3 report does not specify a safety objective with regard to individual or collective risk, which would require level 3 assessment, i.e. an assessment of the consequences of a release for the public and the environment in terms of the impact on public health and the economy.

---

(1) Reference HBR/BBD-H311 of 26 March 1990.
(2) Reference M-238.
(3) Y.G. Gonen - HSK.

The INSAG-3 report also mentions the use of probabilistic analysis in its description of basic principles, where the complementarity of probabilistic and deterministic safety analyses is emphasised with regard to both the definition of safety provisions and the assessment of their merits, and in that of individual principles with regard to the definition of reliability goals for equipment and systems, for example, and the external hazards that need to be taken into account in power plant design.

This brief overview provides me with an opportunity to stress two important points:

- Probabilistic safety studies of specific installations aimed at obtaining an overall assessment of risk, which are currently carried out in many countries (1) and which will be discussed in the last section of this paper, account for merely a portion of the probabilistic analyses used in nuclear safety management;

- Nuclear safety is not based solely on probabilistic analyses; considerable use continues to be made of deterministic methods, where the benefit of past industrial experience make it possible to protect an installation against identified technical hazards.

In discussing both the potential benefits and the limitations of probabilistic safety analyses (which from now on I shall refer to as PSAs) in the design of power plants, I feel that a distinction must be made between the different levels of assessment.

LEVEL 1

With regard to level 1 assessment, risk of core melt, there have been numerous studies made worldwide which have been subjected to detailed critical review and which have thereby acquired considerable credibility. I feel that it should be stressed that these are "major" assessments requiring minutely detailed analysis of not only the design of the plant but also its operation in all configurations covered by the operating specifications, and drawing on reliability data for equipment and operatives which have been subjected to critical review to ensure that they are truly representative of the plant being assessed. As a general rule these PSAs are now "transparent", that is to say they are readily comparable in terms of their basic assumptions, design models and results. The results of such analyses still include a margin of uncertainty, but one that is now relatively small. I would say that in terms of overall probability, if the scope of the analysis is limited to failures and hazards inside the plant, the interval between 5% and 95% in a good PSA is of the order of a decade. I therefore feel that we can confidently claim that we now have a well honed and serviceable tool at our disposal for use not only in individual safety analyses, which I shall be discussing later, but also in overall risk assessment.

---

(1) Since publication of the Rasmussen report in 1975, which still remains the standard reference, examples include: Biblis in the FRG, the five US plants compared in the NUREG-1150 report, Sweden, Switzerland, etc., and the recent publication in France in May 1990 of two level 1 PSAs, one for a 900 MW PWR and the other for a 1300 MW PWR.

I have mentioned my reservations with regard to external hazards. I think that we should await the conclusions of the NUREG-1150 "Peer Review", currently being conducted by a team of international experts under the supervision of H. Kouts, before we can be as confident as I have been in the case of "internal" assessments, particularly with regard to the account taken of earthquakes.

As for the other limitations of level 1 PSAs, my own opinion is that, provided that the team carrying out the analysis has a good knowledge of plant operation under normal and accident conditions, the advances that have been made over the last few years have significantly reduced the levels of uncertainty which are usually cited and which may briefly be outlined as follows:

- Non-exhaustiveness of initiating events; past experience provides us, I feel, with an adequate guarantee of the exhaustiveness of relatively frequent events, i.e. of an order of magnitude of >10-3; and the account taken in deterministic analyses of various rare failure modes makes it highly improbable that a new type of event might produce a risk-dominant sequence;

- The allowance made for common-mode failures; once analysts have made an in-depth investigation of potential interactions between systems, methods based on the use of operating experience are now available today which, in my opinion, produce fairly good results; this is nonetheless one area in which it is essential to pay special attention to all operating events in order to identify any common-mode failures that might have been overlooked. The same is true of accident scenarios; operating experience may reveal unsuspected connections;

- The allowance made for human factors; our experience at EdF has taught us that by means of simulations it is possible to obtain an adequate qualitative and quantitative knowledge of behaviour on the part of both operating and maintenance staff which might initiate or aggravate accident sequences. It should be noted, however, that such tests and collection of data involve a considerable amount of work.

LEVEL 2

Whereas a level 1 PSA of an accident ends once the core is no longer adequately cooled, a level 2 PSA must examine all subsequent phases of the accident until the situation has been definitively stabilised. The analysis must then examine the behaviour of all structures in the installation that will have a role to play in containing radioactive products, particularly the containment itself with all the possible containment by-passes which might occur under operating conditions for which these structures were not designed. Lastly, the analysis must track the pathways of the radioactive products themselves, which, according to the radionuclide concerned, will depend upon their physico-chemical composition and their evolution. In view of the above, it is clear that the conclusions that had drawn with regard to level 1 cannot apply to level 2 analyses:

- The uncertainties involved are much greater and much harder to quantify in that for many of the problems raised, which I shall list below, we are almost entirely dependent on the opinions of experts;

- The models describing physical phenomena often exhibit major short-comings: scenarios can no longer be established by means of the Yes or No branches used for level 1 analyses, since in some cases they must be grouped according to type of phenomenon, thereby eliminating their specificity;

- We lack sufficient data to quantify, and also sometimes to define, certain sequences in sufficient detail; it is therefore difficult to ensure exhaustiveness;

- The analysis must necessarily cover long periods of time, which tends to complicate the procedure without necessarily ensuring that the various operating options available to the operators are properly taken into account.

I feel that the US NUREG-1150 report, together with the critique that will be made of it, will provide a standard international reference with regard to all the above points for some years to come. I do not think that other studies of this kind will be able to produce better results. Indeed, my personal feeling is that we can learn a lot from partial probabilistic analyses designed to make an in-depth investigation of containment behaviour for some of the reference sequences that would seem, a priori, to be the least hypothetical. Although deliberately abandoning any attempt to be exhaustive, it should still be possible to assess certain weak points and, where necessary, put them right.

LEVEL 3

Level 3 analyses address the final consequences. Clearly they start with all the uncertainties covered at level 2 in terms of both probabilities and source terms. In addition, the analysis encompasses problems relating to the release of radionuclides to the environment, the effectiveness of emergency response plans and clean-up programmes, as well as the impact of low doses.

In my opinion, the findings of the NUREG-1150 report, expressed as the annual risk of premature death and delayed cancer, when compared with those of the Rasmussen reference, do not have any significant lessons to offer in terms of safety. With regard to the general public, I fail to see what there is to gain from the fact that for Grand Gulf and Peach Bottom, for example, we have gained three decades on the risk of premature death assessed in 1975, but merely one decade on the risk of cancer!

Contrary to my comments on level 2 analysis, I feel that there is little be gained from proceeding with level 3 PSAs or even partial PSAs. Perhaps might be worth noting that implementation of the ICRP recommendations on tential exposures would in fact require all analyses to be systematically tended to this level.

V. THE ROLE OF PSAs

I would classify the potential applications of a PSA under five different headings:

- Design of new reactors; in my opinion, since a detailed description of such a reactor and its mode of operation would not be available beforehand, PSAs of this kind must still make reference to a reactor already in service; if the reactor is of a radically new design and therefore cannot be linked to an existing model, then to my mind a PSA is unsuitable and a partial assessment would be more appropriate;

- Operation of a reactor already in service; a PSA could be used to identify the risk-dominant sequences in order to provide input for decisions regarding any modifications that need to be made to the installation or the operating, maintenance and periodic testing procedures for the facility;

- As part of the dialogue between the utility operator and the safety authorities with regard to all questions that fall within the competence of the latter; while a PSA should not be used for "policing", it can provide useful clarification in technical discussions;

- As an aid to the development of research programmes, by helping to identify points where a better understanding would be either useful or necessary;

- Where necessary, to lend greater rigour to democratic debate, where it should be used more modestly.

To my mind, before a PSA can properly fulfil its role, it must meet several conditions:

- It must be above reproach in terms of quality and transparency; we must beware of "light-weight" PSAs which have not been subjected to critical appraisal and whose intricacies cannot be readily appreciated by outside experts;

- It must be continuously kept up to date both in terms of data, which can be obtained from databanks containing data relating to reliability, equipment and staff, and in terms of models and scenarios, where use can be made of past experience; this means that, in view of the use made of computer facilities, a PSA must be designed with a view to such updating;

- It must be presented in a user-friendly format so that it can be used by technicians who, although they have clearly been specially trained to use such analyses, cannot be expected to be experts in reliability and probability procedures.

PSAs are already playing an essential role in the field of nuclear safety. I am confident that within a few years their use will become far more widespread, even though tried and trusted deterministic analysis and respect of the principle of defence in depth will undoubtedly remain as important as they are today.

A REGULATORY VIEW ON THE APPLICABILITY OF THE NEW ICRP RECOMMENDATIONS TO NUCLEAR SAFETY ASPECTS

E. Gonzalez Gomez
Consejo de Securidad Nuclear
Madrid, Spain

R. Naegelin
Swiss Federal Nuclear Safety Inspectorate
Würenlingen, Switzerland

A.P.U. Vuorinen
Finnish Centre for Radiation and Nuclear Safety
Helsinki, Finland

ABSTRACT

The new ICRP recommendations will cause several changes in the radiological practice. This paper discusses these changes from the point of regulatory view of nuclear safety. To avoid adverse short-term effects, the new risk estimates should be adopted in radiation protection standards with great care.

The ultimate objective of nuclear safety is to protect people, environment and property against radiological hazards. Improvements in principles and practices developed by the ICRP are important in reaching the primary goal. A severe nuclear accident must be prevented in advance. Every scientific and technical means have to used; optimization is not the solution of the problem.

L'APPLICABILITE DES NOUVELLES RECOMMANDATIONS DE LA CIPR AUX ASPECTS DE SURETE NUCLEAIRE CONSIDEREE DU POINT DE VUE REGLEMENTAIRE

RESUME

Les nouvelles recommandations de la CIPR entraîneront plusieurs modifications dans la pratique radiologique. Dans la présente communication, les auteurs examinent ces changements sous l'angle de la sûreté nucléaire considérée du point de vue réglementaire. Afin d'éviter des effets néfastes à court terme, il convient de prendre de grandes précautions, lors de l'adoption de ces nouvelles estimations des risques dans les normes de radioprotection.

La sûreté nucléaire a pour objectif ultime de protéger les personnes, l'environnement et les biens contre les dangers des rayonnements. Les améliorations des principes et pratiques élaborées par la CIPR revêtent de l'importance pour la réalisation de cet objectif primordial. Il faut prévenir la survenue d'un accident nucléaire sévère. Tous les moyens scientifiques et techniques doivent être mis en oeuvre ; l'optimisation ne constitue pas la solution de ce problème.

1. Introduction

Since the beginning of the wide industrial applications of nuclear
power concerns about the radioactivity impact were a matter of
analysis. Regulatory Nuclear Safety requirements were first established in the appendix A of the 10CFR50 in the U.S.A. As it is
stated there the final aim of the Nuclear Safety Requirements is
to provide adequate protection for the public and the workers from
undue radioactive risks.

The latest document that has undertaken a systematization of the
safety requirements has been the INSAG-3 report "Basic Safety
Principles for Nuclear Power Plants".

The safety of a Nuclear Power Plant (NPP) is achieved by consideration of many criteria for design, construction and operation,
taking into account normal operation and accident conditions.
Figur 2 of the INSAG-3 report gives a presentation of 50 specific
safety principles. All these principles have the purpose to limit
radiation doses to the plant personnel or to the public or to prevent or to mitigate accidents, i.e. potential exposure situations.
Radiation protection is taken care of in siting, design, operation
and emergency measures (yellow).

These principles are not aims for themselves. The final aim is to
protect people, environment and plant. What this graph shows is
the great number of principles, which to a large degree are redundant, resulting in a defense in depth concept for normal operation
and accidents.

The new ICRP recommendations will cause, if followed, several
changes in the radiological practice. To avoid adverse short-term
effects, they should be adopted with great care.

The general task of a nuclear regulatory organization is to verify
and, if necessary, to ensure that nuclear activities are exercised
without undue harm to the society.

The regulatory activities are normally anchored to the legislation
on lower level regulations, where the necessary procedures, frames
and other constraints for regulatory decisions are established.

National practices in various countries differ, but many common
features can be found as well. Especially on areas where nuclear
safety is related to radiation protection the previous recommendations of the ICRP have had a harmonizing effect on practicies.

Many ICRP recommendations have become part of formal regulations
in our countries. This is practicable as long as measurable quantities such as radiation doses are concerned. It does also work
with general principles such as ALARA, when a common practice of
application has been established. Formal regulation would not be
possible however for quantities that cannot be measured or calculated according to an established state of the art and with sufficient accuracy.

The new ICRP recommendation document includes an extensive treatment of the risks of ionizing radiations and proposes some new approaches to be applied in radiation protection in general.

We believe that to apply the recommendations on nuclear safety activities the following issues are of major importance:

- Increased radiation risk estimates (per unit dose) and compensating reduction of dose limits.

- Requirement to regulate the risks of exposures due to accidents by the potential exposures approach.

- Treatment of pre-existing situations.

- Possible legal consequences of these new aspects, especially of point (2).

Let's take these issues one after the other.

2. Risk estimates and dose limits
(Fig. 1: ICRP Recommendations Draft February 1990 Fig. 5, p. C-25)

An essential cornerstone of radiation protection is the knowledge on the relationship between radiation dose and health effects. The uncertainty on this basic issue is still remarkable.

The proposal to increase the risk factors is based on new scientific information but at the same time it is a serious signal about the limitations of the knowledge on this area. The weakness in our basic understanding is by no means creating confidence towards the public.

Since the previous risk estimates it has been required to improve the safety of the general public and the occupational conditions.

Increased radiation risk gives the reason to correct the dose limits (Fig. 2: ICRP Document Table 6, p. 5-16). How this is done, without distorting the harmonized system, is an important issue to be discussed between nuclear safety and radiation protection regulators.

The new values of risk estimates probably rise at least the following additional problems:

- the strong age dependency of the risk is not taken into account in the establishment of the dose limits

- public opinion might be negatively affected

- legal proceedings may become more frequent when persons exposed in the past develop a malignancy.

The last two issues should not concern the ICRP or the regulatory authorities. The politicians, industry and courts will have to resolve them.

Some possible consequences of the new limits might be, that

- the monetary coefficients of the radiation induced harm and risk, used in the justification and optimization procedures,

165

will have to be increaced ($ per unit dose or unit potential dose) - logically in proportion to the change in the risk factors

- some of the previously executed justification and optimization procedures may have to be revised.

An important question is whether the employers of the exposed professionals will be able to cope with the proposed dose limits; respective problems are known in uranium mines, but also for surgeons working with X-rays. As far as the operation of NPPs is concerned, the answer seems to be positive. Difficulties regarding specific tasks in existing facilities might be resolved by the two "outlets" of the recommendations, namely pre-existing situations and "special operations of considerable importance", at least for a transitory period until technical solutions will be found.

Let's illustrate the situation for the operating NPPs by some figures:

- The average individual dose for the staff of NPPs is well below the new limit of 20 mSv/a proposed by ICRP (Fig. 3, from a NEA report published this year: Occupational Dose Control in NPP, figure 2)

  The average doses show a tendency to decrease as may be seen by the development of the collective doses of single NPPs. 10 years ago collective doses of 10 man Sv/a were not extraordinary; today the design objective for advanced reactors lies in the order of 1 man Sv/a.

- While the average annual dose of NPP-personnel is well below 10 mSv/a, there is still a large number of persons exposed to doses above 20 mSv/a. In Switzerland this number has been lowered from 271 persons in 1986 to 75 persons in 1989. Further efforts with respect to automation, remote handling and reduction of contamination will be necessary (Fig. 4).

According to the information presented in the "ICRP Draft Recommendations" the risk of ionizing radiation depends as well on the doses as also on the age of the exposed person. Hence this risk may be lowered not only by reducing the doses, but also by exposing older persons to given doses. The risk of the same dose is more than 3 times smaller for men between 50 and 59 years old as compared to men between 20 and 29. When all age groups between 20 and 60 years are equally exposed to the same incremental risk instead of the same dose, the total risk for the same collective dose is still about 20 % smaller. Nuclear power plants are the most important contributors to the collective occupational radiation dose. On the other hand they have also a well organized control over their personnel. It is suggested therefore that for such cases an option for age dependent dose limits is made available (e.g. by defining a limit for the accumulated dose as a function of age). The option may be constant annual risk increment instead of constant annual dose; it would open an additional way for the optimization of radiation protection, allowing an easier transition to the lower dose limits and a lower total risk involved with the occupational radiation doses (Fig. 1).

Such an alternative option could also keep the old limit of 50 mSv/a, as long as one stays below the admissible age dependent accumulated dose or risk. This would allow not only more flexibility at the same risk, but it would also allow many workers to stay with their given dose history within the new limit, a limit that would have been hurt in retrospective if it just says 20 mSv/a. The change from the old to the new limits would appear less dramatic and quite a number of court cases could be avoided, and this of course without accepting a higher risk.

3. Regulating plant safety by the limitation of potential exposures

(ICRP Doc. § 203 p. 5-18: Individual risk limits:
$8 \cdot 10^{-4}$/a for occupational exposure
$5 \cdot 10^{-5}$/a for public exposure)

Regulatory requirements for nuclear safety are concentrated on avoiding core damage or even core melt, since this is a point where the radiological consequences are large, both for workers and for the public. Estimates about the radiological impact to the public are uncertain. The release is influenced by several factors, such as fission product behaviour in the containment and the containment failure mode, doses to members of the public additionally by the weather category and the effectiveness of off-site emergency measures. Regulation of the potential dose or risk ("back end") hence is problematic and could distract the attention from the main objective, the avoidance of core melt ("front end").

The potential detrimental outside effects of a severe nuclear accident are the direct individual and collective radiation doses to members of the public and the contamination of the environment. While protection of the public by appropriate emergency measures can be accomplished, avoidance or cleanup of contaminated land is not possible or is very difficult. Therefore, limiting only the annual risk for individual members of the public would not address the main problem of severe accidents.

The damage caused by an accident depends on population density and use of land in the environment of the plant. Limitation of dose or risk to individuals does not take into account this dependency as long as there are no large exclusion areas. Severe accidents with release of large amounts of radioactive material must not occur anywhere and most certainly not in densely populated areas. Severe accidents shall be prevented by all means provided by the science and technology; consideration of cost-benefit, i.e. optimization of protection to achieve minimum cost, is not acceptable in such cases. The success of these preventive measures is checked by a plant specific Probabilistic Risk Analysis (PSA) of levels 1 and 2. Acceptable frequencies of core melt probability and of large releases depending on the scope of the analysis are discussed within the regulatory authorities. The state of the art does not yet allow to set formal limits. Pioneering work has been done by the NRC.

Evaluation of the risk to the members of the public can be done by PSA of level 3. Additional uncertainties are introduced and, the approach does not cover the entire area of problems. The result may be used in an informal way for judging the safety of a plant or for making generic decisions with respect to the design or the operation.

A value of $5 \cdot 10^{-5}$ is recommended by the ICRP for the annual risk limit for individuals of the public in potential exposure situations. This figure is relatively high, when compared e.g. with the quantitative risk objectives of the NRC, $0,5 \cdot 10^{-6}$ and $2 \cdot 10^{-6}$ per reactor year (acute and delayed risk respectively) i.e. 20 times lower. Also assuming that the NRC rule is valid for 1 reactor and the ICRP proposal for all reactors at a site, there still would be a considerable difference. If the figure of $5 \cdot 10^{-5}$ is used as sole criterion for the acceptability of severe accidents, it could discourage operators to install useful safety features. If it is used beside other criteria, it will most probably never be the limiting criterion and a lot of work would have no consequences at all.

There are other detrimental effects of severe accidents, e.g. radiation doses to plant and clean-up personnel (Fig. 5).

The financial consequences of a core melt accident are so large, that the government, regulators and owners of nuclear power plants have to include also this aspect in their policy. The result of an optimization therefore depends largely on the scope of activities which are considered and compared. Guidance for the application, and guidance for the exclusion (or non-applicability) of optimization procedures would be necessary.

To cope with severe accidents in nuclear power plants an attempt has been made, after the TMI-accident in 1979, by the USNRC to define Safety Goals. Since then extensive exercices have been carried out in most member countries and much experience has been gained through the trial use of probabilistic and quantitative safety guidelines. The international organisations IAEA, NEA/OECD and CEC have reviewed the activities on the safety goals all over the world. Status reports on this subject are now available:

- Guidelines on the role of probabilistic safety assessment and probabilistic safety criteria in nuclear power plant safety (a safety guide) to be published in Safety Series by the IAEA.

- Consideration of quantitative safety guidelines in member countries, a report by a group of experts of the NEA Committee on the safety of nuclear installations, OECD-NEA, to be published.

- CEC-Summary report on safety objectives in nuclear power plants, EUR 12273 EN, Brussels 1989.

In these reports the two aspects, nuclear safety and radiation protection, are treated in a comprehensive way. The concern of potential exposure, as introduced by the ICRP, is well taken by carefully designed probabilistic safety goals for nuclear power plants.

4. Pre-existing situations

These situations are of two types, static and dynamic, and the differentiation between them is appropriate. They also may involve planned and potential exposures.

Examples of static situations are the radon problem in buildings, the long-term or far-field post-accident situations or one in which backfitting is considered e.g. to reduce accident probabilities. To such situations the ICRP approach of considering the effects of a contemplated action using the justification and optimization procedures, sometimes even without regard to the dose limits, is applicable.

There are other examples where the applicability of this approach is problematic. When backfitting to reduce large potential exposures (the probability of severe accidents and the magnitude of their consequences) is considered, all technically feasible and effective measures should be taken, without recourse to the justification and optimization procedures. Such actions, per definition should be deemed justified.

Emergencies are dynamic situations, which require urgent decisions, especially in their early phases. These decisions can be based only on predetermined plans and action levels. It is questionable whether this specific situation belongs to the general recommendation of paragraph 129, recommending against the application of preset dose limits for pre-existing exposure situations.

5. Possible legal consequences

The higher risk estimates and the requirement to regulate the risks of exposures due to accidents surely raise the frequency of court cases. The increasing number of claims for compensation of loss of health is not directly an issue of nuclear safety. It is a matter of experts of radiation protection, radiation biology and insurance.

There are national practices to handle these cases but little international cooperation up to now. ICRP risk evaluation may help to harmonize the radiological risk asessment. However, the compensation practice is mainly a matter of insurance policies.

The requirement to regulate risks of exposures due to accidents by potential exposures approach may open an unresolvable problem. It is necessary to use probabilistic methods in analyzing the risk of nuclear installations. Probabilistic approach surely can be practically used even in regulation but direct binding probabilistic requirements should be developed with great care. It is also useful, and probably necessary, to use safety goals for potential exposure cases, as it is already done in some countries. However, we believe that the decisions also in the foreseen future are more or less based on deterministic requirements.

6. Some other, specific issues

Tissue weighting coefficients (W_T -s)

ALI-values are changed due to the revision of the W_T -coefficients. ALI-values are important imputs especially into accident assessments, and should be provided as soon as possible. The W_T -s given in the document apply to adults. However, the critical group will usually consist of children. It would be important to provide W_T -s for children too.

Collective potential detriment (e.g. to decide on protective measures)

This quantity is required for the justification and optimization procedures in context with potential exposures. Because it is not defined additional efforts are needed before getting any practical value.

Total societal detriment

For the optimization of safety (i.e. to decide on preventive measures) the total societal detriment or cost should be considered instead of health related detriment alone. It should include, in addition to the value attributed to the health detriment, power replacement, investment loss, land loss and clean-up costs (to mention some major items). This is missing from the draft document.

Non-nuclear related radiation practices

It is appropriate to mention the impact of the proposed ICRP recommendations on the regulation of non-nuclear related radiation practices. In this field accident prevention is the main approach also and is done most effectively by deterministic methods.

7. Conclusions

The conclusions are summarized in 3 groups:

1. General Safety Principles

- The ultimate objective of nuclear safety is to protect people, environment and property against radiological hazards. This has to be taken as a key element in design, construction, operation, dismantling and final disposal of nuclear installations. Improvements in principles and practices developed by ICRP are important in reaching the primary objective.

- Most nuclear safety principles and practices stand on their own as mere industrial safety instruments that take into account the technical, economical and social implications, with radiation protection as a component of the overall decision making.

- Efforts have been made to develop the optimization principle which, if not taken as a rigid mathematical model of decision making, has always been a characteristic of good engineering practice. Improvements have to be made and important results can be obtained through the wide exchange of information. Additional progress is possible by the application of probabilistic risk assessment.

2. Risk of Radiation

- The new risk estimates have to be adopted in radiation protection standards with great care avoiding transient effects.

- All data and relevant studies on the relationship between occupational radiological work and health effects should be utilized.

- There is a need to develop specific recommendations on doses and activities which are below regulatory concern.

- Development of age dependent dose limits should be considered.

3. Severe accidents, potential exposure situations

- Several detrimental effects of severe nuclear accidents are possible: doses of the operating or cleanup personnel and financial losses for the operating organization, doses to the public and contamination of the environment.

- Limitation of the doses or the risks of the individual members of the public does not address the whole problem area of severe accidents. Moreover, it cannot be applied today in a strict formal way, as is the practice with other ICRP recommendations on numerical limits.

- Formal requirements with respect to severe accidents are connected to the design and the operation of the nuclear installations. These requirements have a deterministic character. Probabilistic safety goals are used as guides depending on the scope of the analysis. They may help, in connection with probabilistic analysis, to identify the weak points of a plant and to judge the probability and the size of potential source terms.

Acknowledgment

The authors of this paper express their appreciation to Mr. Y.G. Gonen who made studies on the subject. Results of his studies are utilized in this paper.

Figure 1

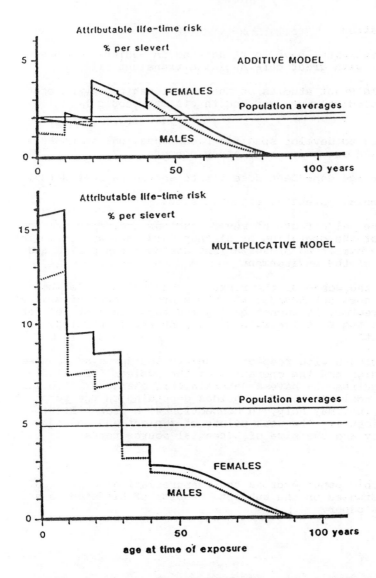

The attributable lifetime risk from a single small
dose at various ages at the time of exposure, assuming a DDREF of 2. The discontinuities are the result
of the use of constant annual values for the primary
risk coefficients within 10-year age intervals (cf.
Table 1). The higher risk for the youngest age group
will not be expressed until late in life.

Figure 2

RECOMMENDED DOSE LIMITS [1]

Application	Dose limit	
	Occupational	Public
Effectance	100 mSv in 5 years 50 mSv in any 1 year	1 mSv per year, averaged over any 5 consecutive years
Annual dose equivalent in		
the lens of the eye	150 mSv	15 mSv
the skin (100 cm^2)[2]	500 mSv	50 mSv
the hands	500 mSv	50 mSv
Mean dose equivalent to the fetus[3]		5 mSv after diagnosis

NOTE 1.   The limits apply to the sum of the doses from external exposure in the specified period and the 50-year committed dose (70-year for public exposure) from intakes in the same period.

NOTE 2.   See Paragraph 173 for details.

NOTE 3.   This value relates to the occupational exposure of pregnant women, which is discussed in Section 5.3.3.

Figure 3

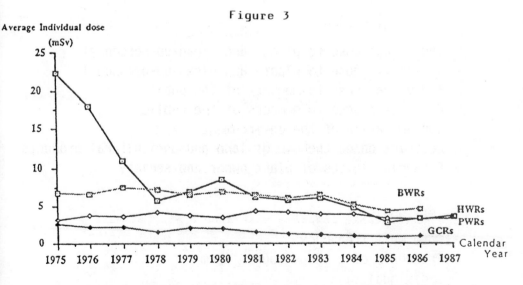

Average Annual Individual Doses for Personnel Working in the Main Reactor Types in the OECD Member Countries (except France)

Figure 4
Occupational doses in Swiss Nuclear Power Plants

Year	Number of persons with doses > 20 mSv	Maximum dose (mSv)	Average dose (mSv)
1986	271	46	9.2
1987	122	40	6.8
1988	87	38	6.7
1989	75	34	6.4

Figure 5

Detrimental Effects of Severe Accidents and Safety Goals

Detrimental Effects:

- Individual dose to plant- and clean-up-personnel
- Collective dose to plant- and clean-up-personnel
- Individual dose to members of the public
- Collective dose to members of the public
- Contamination of the environment
- Restriction on the use of land and agricultural products
- Financial losses of plant owner and society

Safety Goals:

- Core melt frequency
- Frequency of a large source term
- Individual risk
- Collective risk
- Contaminated area

IMPACT OF REDUCED RADIATION EXPOSURE LIMITS ON NUCLEAR PLANT OPERATIONS

C.W. Fay
Wisconsin Electric Power Company
Milwaukee, Wisconsin, United States

C.J. Wood
Electric Power Research Institute
Palo Alto, California, United States

ABSTRACT

This paper assesses the impact of lower exposure limits on operations and maintenance activities in the context of the decreasing trend in exposures at U.S. nuclear power plants. The number of workers with annual doses exceeding 2 Rem (20 mSv) decreased by 40% in 1989 compared to 1988, but some key experienced workers, particularly skilled labor contractors, would be impacted by possible future limits on lifetime doses. Mitigating actions implemented by utilities are discussed, including dose leveling, enhanced training programs and radiation field control technology.

INCIDENCES D'UN ABAISSEMENT DES LIMITES DE RADIOEXPOSITION SUR L'EXPLOITATION DES INSTALLATIONS NUCLEAIRES

RESUME

La présente communication est consacrée à une évaluation des incidences d'un abaissement des limites d'exposition sur l'exploitation et les activités de maintenance dans le contexte de la tendance à la baisse des expositions dans les centrales nucléaires des Etats-Unis. Le nombre des travailleurs ayant reçu des doses annuelles supérieures à 2 rem (20 mSv) a diminué de 40% en 1989 par rapport à 1988, mais certains travailleurs clés expérimentés, en particulier des sous-traitants qualifiés, seront touchés par les éventuelles limites futures applicables aux doses sur la durée de vie. Les actions correctrices mises en oeuvre par les compagnies d'électricité sont examinées, notamment l'écrêtement des doses, les programmes de formation renforcée et les techniques de réduction des champs de rayonnements.

INTRODUCTION

The federal standards for radiation protection, as contained in Title 10 Code of Federal Regulations Part 20 (10 CFR 20), state that the maximum occupational exposure limit to the whole body for a worker with a complete record of his/her exposure on file, is 3 rem/quarter,* with a lifetime limit of 5(N-18), where N is the worker's age in years. In theory this allowed for an annual exposure of 12 rem. In the past some workers have approached this annual limit [1], but now all utilities adhere to the INPO recommended guidelined of 5 rem/year. A recent report by the National Council on Radiation Protection and Measurements (NCRP) [2] recommends that "... the numerical value of the individual worker's lifetime effective dose equivalent in tens of mSv (rem) does not exceed the value of his or her age in years." A new draft revision to the regulation by the NRC will limit exposures to 5 rem per year in line with the INPO guideline, international recommendations ICRP 26, 30 and 33, earlier NCRP recommendations, and current practice.

Recent reassessement of the health risks of ionizing radiation (for instance the BEIR V study [3] have resulted in preliminary recommendations to reduce individual exposure limits. A draft report of the International Commission on Radiological Protection suggests an annual limit averaging 2 rem/year (20 mSv/year).

For some years, U.S. utilities have been committed to maintaining exposures as low as reasonably achievable (ALARA). Recognizing the potential for more restrictive exposure limits, U.S. utilities have aggressively reduced individual and cumulative exposures [4].

RADIATION EXPOSURE TRENDS

Occupational exposures have decreased every year since the early 1980s. 1989 saw one of the greatest drops, from 345 to 290 man-rem (cSv) at PWR plants and from 510 to 440 man-rem (cSv) at BWRs (Figure 1).

Since 1982, the annual electric generation from U.S. nuclear plants has doubled, whereas the total radiation exposure has decreased 20% (Figure 2). The ratio of man-rems to MW.years, which is a measure of the industry's effectiveness in radiation protection, has dropped 50% in the past five years.

This outstanding success story of the nuclear industry has passed largely unnoticed in the current media debates about radiation health risks. Recognizing that exposure limits could potentially become more restrictive, the utility industry adopted a proactive approach several years ago. Figure 3 shows that, based on 1982 plant averages, total exposures would have increased from 55,000 man-rems (550 Sv) in 1982 to 75,000 (750 Sv) in 1989 as more plants came on line. In fact, the total decreased to 39,000 (390 Sv) over this period.

Figure 4 shows the cumulative exposure plotted against cumulative energy produced since 1973. Apart from a decreasing trend in the late 1970s as many new

---

* 1 rem = 10 mSv

plants came on line, and an increasing trend after the Three Mile Island accident, the graph was approximately linear until 1984, but has exhibited a favorable decreasing trend since then.

COMPARISONS WITH EUROPE AND JAPAN

Figures 5 and 6 show that the large difference between U.S. plant exposures and those of other major industrial nations with substantial LWR programs decreased significantly through the 1980s. Average exposures have declined in most countries, and although U.S. exposures are still higher, both BWR and PWR plant averages are in the same range as most other countries 3-5 years ago.

For all countries it is generally true that older plants have higher exposures than later plants. This is because, like children, long-term behavior is heavily influenced by what happens in early life. With the exception of Swedish BWRs, little attention was paid to avoiding in-core cobalt sources or to tight water chemistry control in the 1970s, resulting in high cobalt-60 inventories, which can only be reduced slowly by subsequent actions. Even so, exposures are declining at pre-1980 plants, but as there are relatively more of these older plants in the United States, average plant exposures tend to be distorted by this age factor, particularly by comparison with France, where the average age of plants is only about 6 years.

West German PWR plant averages benefit from some extremely low exposures reported by the most modern plants, whose designs post date those of the latest U.S. plants, which were all designed before 1977. The latest German PWRs have largely eliminated cobalt materials, which has greatly reduced formation of the gamma ray emitting cobalt-60 isotope and hence radiation field buildup. The latest BWRs in Japan are also showing the benefits of recent design: the installation of large feedwater clean-up systems and the capability to adjust water chemistry within tight specifications has resulted in exposures a factor of 10 lower than in early Japanese plants.

FACTORS INFLUENCING THE EXPOSURE TRENDS

In the early 1980s, post TMI-2 plant modifications and several PWR steam generator and BWR piping replacement projects contributed to increasing exposures by 40% or more. The reduction in the number of these major modification or repair projects and an ability to perform the projects more effectively have resulted in decreased U.S. utility exposure totals. More recently, improved water chemistry, cobalt source replacement and use of chemical decontamination have reduced radiation fields. Figure 7, depicting exposures incurred in steam generator replacements, illustrates the general trend of reducing exposures associated with major repair work.

INDIVIDUAL WORKER EXPOSURES

The number of workers receiving doses approaching the 5 rem (50 mSv) limit has decreased markedly in the past few years (Figure 8). In 1989, there were no exposures in excess of 5 rem and the number of workers receiving more than 2 rem

(20 mSv) decreased from 2100 to 1320, a reduction of 40% in one year alone. This illustrates the success of actions by utilities to reduced individual and cumulative exposures.

IMPACT OF REDUCED EXPOSURE LIMITS

Several factors may be affected by reduced exposure limits, such as:

- Restriction in use of experienced workers on high dose jobs,

- Contractors' key workers will be severely restricted,

- High dose work may require use of more crews, to limit individual exposures,

- Reduced flexibility for job planning, increased number of workers and health physics coverage, and extended outage time may result.

The potential consequences of such restrictions may be an increase in total number of workers exposed, and hence in total radiation exposure. Inefficiencies are inherent with larger work forces. Although individual exposures are reduced, the inherent inefficiencies may result in increased cumulative exposures. Increased outage time for special maintenance work may be another consequence, since more crew changes will result in delays for jobs on critical path. These factors lead in turn to increased operations and maintenance costs—already rising at >10% per year. O&M costs are of particular concern to many utilities, since the economic margin of nuclear power over other generation sources has already been significantly eroded.

EPRI has attempted to examine the impact of lower exposure limits in more detail, through a study by Deltete et al. [5], at a utility which operates two PWR units and one BWR unit. Part of this study focused on so-called "critical workers", that is workers that received more than 1 rem (10 mSv) exposure during 1989. 1989 was a fairly typical year for this utility, as one PWR had a 3-month outage for refueling and maintenance work, with an overall load factor of 69%. The second PWR unit had no refueling outage and achieved 85% load factor. The BWR plant had a 2-month refueling outage and achieved 77% load factor over the year. The PWR plants totalled 340 rem (3.4 Sv) and the BWR 460 rem (4.6 Sv).

The 190 "critical workers" received 33% of the total dose in 1989, even though they represented only 5% of the monitored staff at the plants. The 190 workers were comprised of 80 station staff and 110 vendor (or contractor) staff. For this study, the 5-year dose records were examined for these critical workers, in order to determine the percentage of their work that was performed when their exposures for that specific year had exceeded 1, 1.5, 2, 2.5, 3 rem (cSv). The results were quite different for the station staff and the vendor staff, as shown in Figure 9.

It can be seen that the impact on the station staff is minimal if the annual dose limit is 2 rem (20 mSv) or above, but becomes significant for lower limits. For vendor staff, however, the effect is significant even at 2.5 rem (25 mSv). This study did not consider the impact of lifetime exposure limits,which could further constrain activities of these workers.

One of the main conclusions of this study is that extra resources may have to be provided for training of workers to expand the skilled labor talent base. This is important in view of utilities' commitment that safety must not be impaired through the use of inexperienced workers.

Other mitigating actions include enhanced job planning to minimize exposures of key workers. Almost all utilities have already initiated procedures for controlling lifetime exposures.

TECHNOLOGY DEVELOPMENTS

In addition to enhanced outage planning, implementation of advanced radiation control technology is being accelerated. Examples of techniques to reduce exposures during major maintenance work include robotic devices, and the use of full-system chemical decontamination technology [6]. Partial system decontamination (BWR recirculation piping systems, pumps and reactor water cleanup systems; PWR steam generator channel heads, pumps and heat exchangers) has been a major factor in reducing exposures in the past seven years. Now, several utilities are collaborating with EPRI to provide the qualification work necessary to extend the technology to include the reactor vessel. The qualification programs (BWR and PWR) will be completed early in 1991.

In the longer term, advanced water chemistries will play an increasingly important role as radiation fields are gradually decreased. Examples are zinc injection passivation in BWRs and elevated pH operation in PWRs [7].

CONCLUSIONS

U.S. utilities have been proactive in their attempts to reduce individual and cumulative radiation exposures. These reductions have resulted from the implementation of enhanced technologies and management controls. Although efforts are continuing to further reduce exposures, to reduce individual exposure limit will likely present additional challenges to the U.S. nuclear utility industry. These challenges will be to conform to the lowered exposure limit while minimizing the economic consequences and without compromising the safe operation of the nuclear facilities. Our conclusions are as follows:

- A proactive approach by U.S. utilities has resulted in halving exposures over 5-year period.

- The main problem for U.S. utilities concerns key workers (utility staff and contractors) with high lifetime exposures.

- Increased emphasis on job planning, radiation field reduction technology and use of robotic devices will be required.

- Potential consequences of reduced individual exposure limits include increased outage time, increased O&M costs, increased training to expand labor pool, but safety will not be compromised.

REFERENCES

1. Brooks, B. G. : "Occupational Radiation Exposure at Commercial Nuclear Power Reactors, NUREG 9713, U.S. Nuclear Regulatory Commission, Washington, D. C., 1989.

2. "Recommendations on Limits for Exposure to Ionizing Radiation," NCRP Report No. 91, June 1987, Chapter 8, 25.

3. BEIR V Report : "Health Effects of Exposure to Low Levels of Ionizing Radiation," National Research Council, Washington, D. C. 1990.

4. LeSurf, J. E. : "Implications of Possible Reduction in Radiation Exposure Limits," EPRI Report NP-6291, March 1989.

5. Deltete, C. P. : "Scoping Study: RP/HP Exposure Characterization," EPRI Report to be published.

6. Wood, C. J. : "Review of Application of Chemical Decontamination Technology in the United States," Progress in Nuclear Energy, 23, No. 1, 1990, 35-80.

7. Wood, C. J. : "How the Water Chemistry Revolution is Reducing Exposures," Nuclear Engineering International, February 1990, 27-30.

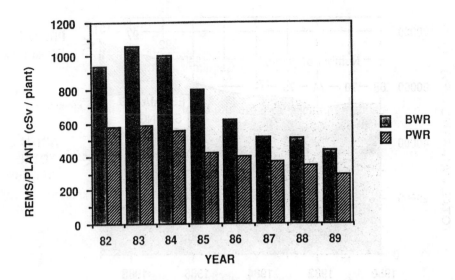

Figure 1.   U.S. Plant Exposures

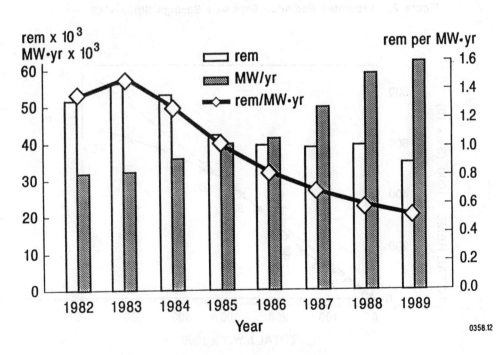

0358.12

Figure 2.   Radiation Exposure and Power Generated at
U.S. Nuclear Power Plants

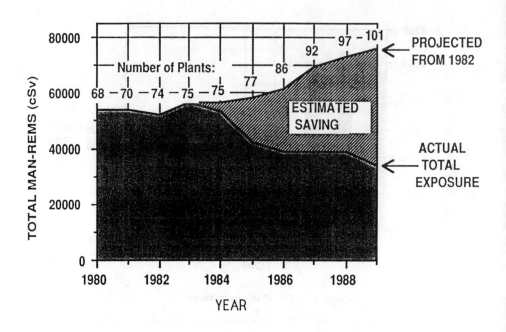

Figure 3. Estimated Radiation Exposure Savings Since 1982

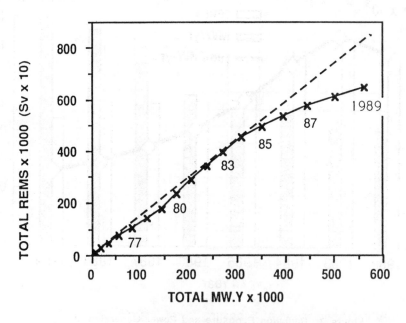

**Figure 4. Cumulative Radiation Exposure as a Function
of Electric Power Generation Since 1972**

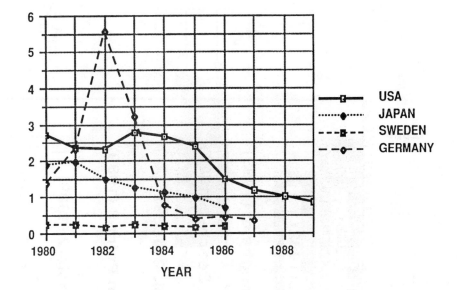

Figure 5. Comparison of Exposure/Power Generation
Ratio for BWRs

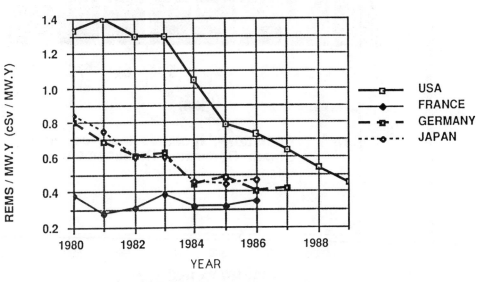

Figure 6. Comparison of Exposure/Power Generation
Ratio for PWRs

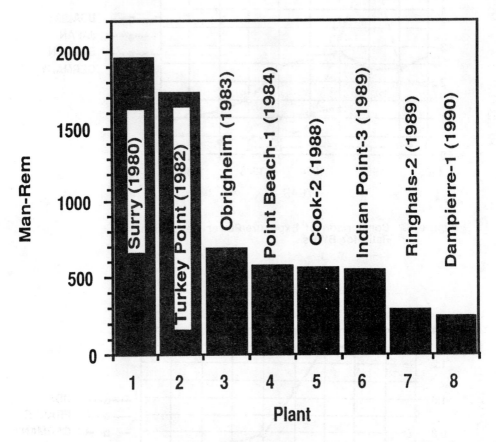

Figure 7. Radiation Exposures Incurred During Steam Generator Replacement

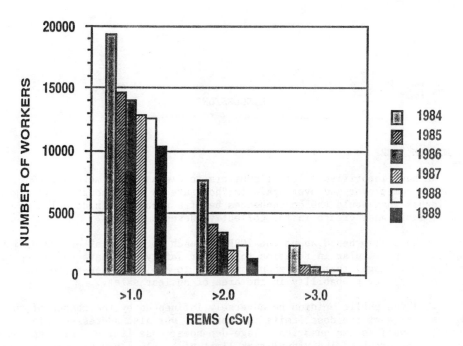

Figure 8. Dose Distribution of Workers at U.S. Nuclear Power Plants

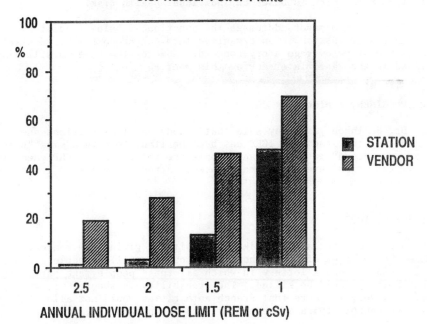

Figure 9. Amount of Work Carried Out by Critical Workers That Would Be Affected by A Reduction in Dose Limit

DISCUSSIONS

D. BENINSON, ICRP

We have identified a lot of subjects to discuss this morning. One question which comes over and over again is "how safe is safe enough". We might also discuss what would the consequences be of a reduction in dose limits. Would it imply an increase in the collective doses?

We have also heard about the ICRP approach to potential exposure situations and in particular in relation to nuclear safety. We have had a suggestion from Dr. Tanguy that ICRP should not try to apply the new approach to events of very low probability in the area of nuclear safety.

Will the public opinion be adversely influenced by the change of the numerical value of the dose limits for workers, was also addressed this morning, as well as the question of age-dependent dose limits. Also addressed today was the proposed individual risk limit of $5 \cdot 10^{-5}$ per year in particular in relation to the ambition levels in the nuclear field, which seem to be higher than that which is based on individual health risk.

Finally, one paper addresses the question of below regulatory concern. Let's run the discussion in an organised form. I propose that we start with the question of below regulatory concern. Some of these items will also be discussed in the Panel later during this meeting.

A.P.U. VUORINEN, Finland

Dr. Beninson yesterday said that nobody has taken actions due to a dose of 1 mSv. At the same time ICRP has been hesitant to give advise "below regulatory concern" matters. Many people are referring to ICRP when stressing that all radiation, even very small doses, introduces a risk. Isn't there a contradiction?

D. BENINSON, ICRP

One cannot set a value like 1 mSv without giving the context. For example, in the area of Paris there are about 10 million people and suppose that I would propose a lottery in which all these people take part. The winner of this lottery would be killed. The probability is about 10^{-7}, which is very small. However, I'm sure that French authorities would not allow this experiment. You cannot separate the context and the probability figure.

The NEA and IAEA have published levels for exemption from regulatory control based on both individual dose and collective dose. If the two things are met, then things could be exempted. This is in perfect agreement with ICRP recommendations.

H. KOUTS, United States

I believe a real problem exists relative to words used by ICRP. They are recognised as saying all radiation is harmful. This is inconsistent with setting a value below which no regulatory concern would be needed. ICRP should state that there are values of dose below which they have no concern.

R.H. CLARKE, United Kingdom

I think Dr. Kouts is talking about a "dose cut-off" when he quotes ICRP as saying that any dose has an associated risk, no matter how small. He and the safety community want to say that below some value of dose there is no risk. Whereas ICRP, when talking about exemption, means a practice can be exempt from regulatory concern even though there are doses and risks to those exposed.

G. NASCHI, Italy

The actual problem is on the practical application of ICRP philosophy especially in an emergency situation. For normal operations of nuclear power plants there are no practical problems to apply the recommendations of ICRP. For emergency situations on the contrary we have a lot of problems. Yesterday for example, Dr. Beninson defined the intervention level for milk used immediately after Chernobyl, 370 Bq per litre, as absurd considering the extremely low risk involved. However, if a similar accident would happen again, we would have the same confused situation, because everybody will interpret the ICRP philosophy in different ways.

The regulations need practical guidance, otherwise the situation will not improve in an international context. Finally, please remember that regulations also have a task to aid legislators in the field of nuclear energy, and there we have to advise on numbers!

These are the practical problems we have.

D. BENINSON, ICRP

We already discussed the question of dose limits versus intervention levels yesterday. The advice from the ICRP is that in an emergency situation you should not use dose limits but intervention levels. It is very difficult to advise on a number for intervention level which is unanimously applicable.

P.Y. TANGUY, France

Les recommandations que publiera la CIPR à la fin de l'année seront présentées comme une conséquence de ce que le risque est trois fois plus important qu'on ne le pensait.

Devons-nous attendre une réduction d'un facteur 3 sur les niveaux d'intervention sur les produits alimentaires, tels que fixés par la Communauté Européenne ?

D. BENINSON, ICRP

No, there is not a one to one correlation between those. What has increased by a factor of about 3 is the probability of attributable death, but
not for example the mean life-time lost.

A. GONZALEZ, IAEA

The discussion of the latter issues emphasizes the importance of clarity
in the vocabulary. We discuss the concept of below regulatory concern (BRC),
an expression that is obscure in regard to the scenario to which it applies.
It is not clear whether it is a dose BRC or a given source or practice BRC.
Depending of our understanding of BRC, our reaction will be different. Inci-
dentally, NEA and IAEA have used the expression exemption and exclusion from
the regulatory system (of notification, registration and licensing) of certain
sources and practices, account taken - interalia - of the individual and
collective doses they deliver.

We have also discussed the applicability of levels rather than of limits
(in emergency situations) perhaps having a different understanding of the concepts of level and limit.

Finally, a question was raised on whether an increase in the risk factor
will necessarily lead to a proportional increase in radiation protection quantities such as intervention levels in food. The negative answer is supported
by the fact that, although the risk has changed, the (multiattribute) detriment
does not necessarily do so. Again, the use of a word for meaning different
concepts is the cause of confusion.

T. TOBIOKA, Japan

The new ICRP recommendations will be kind of a Bible after its publication. However, real implementation of its philosophy to regulation is not easy
due to the use of different "language", vocabulary etc.

We appreciate the NEA organizing this meeting; and I would like to ask
you under another auspice to organize a rather small meeting, to discuss how to
transfer the ICRP philosophy into practical regulation to benefit all countries, and try to identify the real problems of its application.

K.B. STADIE, NEA

I will address this remark in the panel discussion later today.

D. BENINSON, ICRP

I agree with Mr. Tobioka. The language of the ICRP recommendations is
not suitable for regulation and need to be translated into regulatory terms for
those who wish to use the ICRP recommendation in legislation.

Now to the next item. The corresponding risk to the dose limit for members of the public is of the order of $5 \cdot 10^{-5}$ per year. Would somebody like to comment on this figure.

L.G. HÖGBERG, Sweden

Dr. Beninson asked for comments on the $5 \cdot 10^{-5}$ per year figure for increase in individual mortality risk due to potential exposure situations, the figure appearing high in comparison with values derived from basic safety objectives for nuclear power plants such as those proposed in INSAG-3.

To me, the reason for the difference is quite simple: the nuclear safety objective has been arrived at by nuclear safety experts and regulators, taking into account a number of social and economic factors in addition to individual mortality risk. Those factors, combined with state-of-art in nuclear safety technology, turned out to produce lower probabilities as reasonable safety objectives than the ICRP individual health risk concept, but this is not in conflict with ICRP concepts. Indeed, they can be seen as a result of a multi-attribute analysis, although not using the formal models discussed by the ICRP.

S. BENASSAI, Italy

I would like to comment on the $5 \cdot 10^{-5}$ as risk limit. While it can be possible to verify a risk limit for a single event, that is not possible for all the events that can happen in a nuclear power plant. Many events are considered in safety analysis and many more can be envisaged: for such a reason we cannot evaluate the cumulative risk. Given that, the definition of a risk limit is quite useless.

A. GONZALEZ, IAEA

I fully agree with Dr. Benassai that in practice you do not assess individual-related risks but rather source-related risks or, better, scenario-related risks. For that reason in a regulatory context you need criterion curves of the type suggested in our paper; i.e. expressing scenario-related probability limits (as function of dose, as appropriate). Moreover, it should be emphasized that the regulatory compliance with such limits will be ensured (as in normal situations) through derived quantities such as reliability control and operational procedures control. Notwithstanding the above, the individual-related risk limits still have the merit of providing a basic conceptual support to the scenario-related limits, which should be a fraction of the individual-related limits, and the derived quantities.

D. BENINSON, ICRP

The next issue concerns the redution of the occupational dose limit and its consequences. Are we causing an increase in the collective dose?

R.J. BERRY, United Kingdom

I will be dealing with this question in respect of the fuel cycle this afternoon. Suffice it to say that, so far, reductions in worker doses have not yet been accompanied by increases in collective dose. But this has only been accomplished by expenditure of resources on changes in working procedure and man-management. The question of importance is whether this expenditure of resources is justified. At present occupational exposure levels, do we meet our aspirations for a sufficiently low level of radiation and related detriment; in other words, is our present practice safe enough?

A.P.U. VUORINEN, Finland

Considering the risk limit of $5 \cdot 10^{-5}$ per year it is useful to compare that to some other risk figures. That is especially important when developing general safety goals.

The use of electricity is an area of high level of safety; it implies a risk level between 10^{-6} and 10^{-7} per year for the general public. However, the occupational risk is between 10^{-4} and 10^{-5}, i.e. close to $5 \cdot 10^{-5}$. The occupational risk in general of acute death in Finland is about $1.6 \cdot 10^{-5}$, of which 60 per cent is due to traffic accidents when travelling between the home and the work. We should put the nuclear risk (including mining, power plant operation etc.) in perspective when discussing risk limits.

P.Y. TANGUY, France

Je vois deux attitudes différentes dans les propositions de la CIRP : pour la question "de minimis", elles laissent les autorités réglementaires fixer les niveaux en s'inspirant de leurs recommandations ; pour les limites, elles fixent la limite annuelle, 20 mSv par an (alors que le projet de février 1990 donnait 50 mSv).

Or on sait que les doses réellement reçues par les individus résultent avant tout de la mise en oeuvre plus ou moins efficace du principe ALARA. On peut craindre qu'un abaissement radical des limites réduise le champ d'intervention des responsables sur les sites, qui doivent, en bons managers, trouver le meilleur compromis entre exigences et contraintes opposées. Au cours des dernières années, comme l'a montré M. Fay, les doses réellement reçues ont décru considérablement, ce qui prouve l'efficacité des responsables. On peut craindre qu'une réglementation insuffisamment flexible n'aboutisse à un effet opposé.

S.B. PRÊTRE, Switzerland

Par le passé, des champs de radiation inférieurs à environ 10 µSv/h (1 mrem/h) étaient considérés comme négligeables dans la pratique des zones contrôlées. La plupart des instruments de mesure étaient adaptés à cette attitude : ils ne commençaient à mesurer qu'aux environs de 10 µSv/h (1 mrem/h).

Avec une limite de dose située vers 20 mSv par an, on est obligé de prendre en considération aussi des champs de radiation inférieurs à 10 µSv/h, et d'utiliser des instruments plus sensibles.

Cela aura pour conséquence de diminuer aussi les doses collectives. Donc : la réduction de la limite de dose individuelle exigera de modifier les vieilles attitudes vis-à-vis des faibles débits de dose.

B. MICHAUD, Switzerland

Je voudrais revenir sur la question de la dose limite pour les travailleurs. La ICRP propose 20 mSv par an, avec souplesse (flexibilité) possible. Je vois des problèmes au niveau de la législation et de l'application pratique du principe de souplesse qui est préconisé. En effet, si la limite de 20 mSv par an devient une limite légale avec des conséquences pénales en cas de dépassement, comment appliquer le principe de flexibilité qui permettrait un dépassement planifié de la limite ? Il s'agira de déterminer dans quelles circonstances un dépassement de la limite de dose est justifié ou non. Je vois des problèmes. Quelle est l'opinion du président (qui est en même temps Président de la CIPR) à ce sujet ?

D. BENINSON, ICRP

The ICRP recommendations are not meant to be regulations. I would never advise to put the dose limits in law. In my country, for example, the law is simply referring to regulations issued by the competent authorities.

R.E. CUNNINGHAM, United States

The ICRP intended only to provide some flexibility with its 5 year average for occupational dose limits. Averaging would present practical management problems. However, a regulatory agency - or other standards setting organization - could simply set the limit at 20 mSv per year. This would be within the ICRP recommendation. If there is a need, however, regulatory agencies will find a way to regulate the averaging.

R.H. CLARKE, United Kingdom

I wanted to raise four points with regard to the proposal in the paper of Gonzalez Gomez, Naegelin and Vuorinen for age dependent dose limits: social, managerial, risk limits and design.

First, can you live with the social problems of different dose limits for different ages: what will Trade Unions' attitudes be?

Second, can you manage a workforce so that workers side by side doing the same job have different dose limits?

Third, in risk terms, you propose higher doses for 50 year olds - say 50 mSv per year. This can only be done if you reduce the risk accumulated by the 50 year olds in their previous work. You would need to limit doses to 20 and 30 year olds at a level of 5-10 mSv per year. Is this what you want to do?

Finally, how do you design a plant for a working environment dose rate if differently aged workers will be exposed in that environment?

R. NAEGELIN, Switzerland

What we are saying in our paper is that by applying a constant annual risk increment limit - allowing different doses at different ages - instead of a constant annual dose limit, a given collective dose would result in a lower collective risk. The question is whether a regulation allowing an option, e.g. with a maximum annual dose of 50 mSv but not more than an age dependent limit below the line (age-18) 20 mSv, would be compatible with the ICRP recommendation.

J.O. SNIHS, Sweden

In Sweden the annual limit for workers is 50 mSv. In addition to that we say that the accumulated dose at age 30 should not exceed 180 mSv and the accumulated life time dose should not exceed 700 mSv. This mean 15 mSv per year in average and that will certainly be the practical constraint in the optimisation of protection.

A.P.U. VUORINEN, Finland

To Dr. Clarke. The remark in our paper on age dependency of radiation risks follows the ICRP observation of a rather strong such dependency. We, therefore, miss a statement from the ICRP on the implications of this fact as far as dose limits are concerned. It was not a proposal from our side but an observation and a matter for discussion.

L.G. BENGTSSON, Sweden

When considering the differentiation of limits to account for varying individual sensitivity, it is useful to consider the lessons from other instances when this has been tried. In Sweden, we have discussed many such cases:

- Smokers are more sensitive to lung cancer from radon than non-smokers.

- Women have different cancer risks to low LET radiation than men.

- Children may be more susceptible to lung cancers from radon than adults.

- Individuals with light skin are more prone to develop skin cancers from ultraviolet radiation than others.

In all these cases, in the end the decision has been not to introduce special limits for the sensitive individuals. This would be overdoing it in view of the relatively low risk levels involved. I hope that the same conclusion will be reached for the ICRP dose limits.

L. FITOUSSI, France

Je souhaite revenir à la question relative à la flexibilité des limites de doses qui a été soulevée par plusieurs participants et qui est prévue dans les recommandations de la CIPR comme vous l'avez souligné, vous-même M. le Président, laissant les modalités d'application à la réglementation. Après la publication des nouvelles recommandations CIRP, les organisations internationales (AIEA, AEN, CCE ...) vont établir les nouvelles normes fondamentales de sécurité. Si ces nouvelles normes recommandent de maintenir la limite de 50 mSv par an avec les restrictions de maintenir les doses moyennes annuelles au-dessous de 20 mSv et les doses cumulées sur 5 ans au-dessous de 100 mSv, est-ce que cette modalité d'application respecte néanmoins l'esprit des nouvelles recommandations CIPR ?

D. BENINSON, ICRP

Yes, that would be one possibility of writing a regulatory text.

Session 4

ACHIEVEMENT OF RADIATION PROTECTION
AND NUCLEAR SAFETY OBJECTIVES
IN THE REGULATORY AND TECHNICAL PRACTICE (Contd.)

Séance 4

REALISATION DES OBJECTIFS DE LA RADIOPROTECTION
ET DE LA SURETE NUCLEAIRE DANS LA PRATIQUE
AU PLAN REGLEMENTAIRE ET TECHNIQUE (Suite)

Chairman - Président

H. KOUTS
(United States)

APPLICABILITE DES NOUVELLES RECOMMANDATIONS DE LA CIPR AUX ASPECTS DE SURETE NUCLEAIRE : LE POINT DE VUE D'UN EXPLOITANT D'INSTALLATION LIEE AU CYCLE DU COMBUSTIBLE NUCLEAIRE

G.T. Sheppard, R.J. Berry
British Nuclear Fuels plc, Risley, Warrington, Cheshire WA3 6AS
United Kingdom

P. Henry
COGEMA, B.P. n°4, 78141 Velizy-Villacoublay Cedex, France

RESUME

La première partie de la communication a trait aux recommandations de la CIPR et formule des commentaires généraux. Les deux parties suivantes visent les répercussions que les nouvelles recommandations auront sur la conception et le fonctionnement des installations liées au cycle du combustible nucléaire.

Les conclusions suivantes sont tirées :

i)   Les réductions des limites de dose annuelles proposées dans les nouvelles recommandations de la CIPR peuvent être prises en compte dans la majorité des installations modernes liées au cycle du combustible nucléaire, et sont moins rigoureuses que les actuels objectifs de conception.

ii)   Il existe des différences notables au plan national entre le Royaume-Uni et la France dans la capacité d'intégrer les nouvelles recommandations de la CIPR relatives à la limitation du risque imputable aux situations d'exposition potentielle (accidents). Du point de vue français, il n'est pas réaliste de fonder la sûreté des installations industrielles sur des limites de risque exprimées en termes de probabilité de décès, alors que l'expérience du Royaume-Uni largement acquise dans les industries chimiques et autres, de même que dans l'industrie nucléaire, laisse penser que, dans le cas des installations liées au cycle du combustible nucléaire, il ne sera pas difficile de démontrer que les limites de risque actuellement proposées par la CIPR pourront être respectées.

iii)   Il est peu probable que l'exploitation des mines d'uranium en souterrain, aussi perfectionnée soit-elle, puisse être assurée dans les nouvelles limites de dose recommandées par la CIPR.

A NUCLEAR FUEL CYCLE OPERATOR'S POINT OF VIEW ON THE APPLICABILITY OF THE NEW ICRP RECOMMENDATIONS TO ASPECTS OF NUCLEAR SAFETY

G.T. Sheppard, R.J. Berry
British Nuclear Fuels plc, Risley, Warrington, Cheshire WA3 6AS
United Kingdom

P. Henry
COGEMA, B.P. n°4, 78141 Velizy-Villacoublay Cedex, France

ABSTRACT

1. Summary

The first section of the Paper deals with the ICRP recommendations, and makes general comments. The next two sections deal with the implications of the new recommendations on design and operation of Nuclear Fuel Cycle Facilities.

The following conclusions are drawn:-

i. The reductions in annual dose limits proposed in the new ICRP recommendations can be accommodated in the majority of modern nuclear fuel cycle facilities, and are less stringent than current design objectives.

ii. There are significant national differences between the UK and France in ability to accommodate the new ICRP recommendations on risk limitation from potential exposure situations (accidents). The French view is that it is not realistic to base the safety of industrial installations on risk limits expressed in terms of probability of deaths, while UK experience largely gained in the chemical and other industries, as well as in the nuclear industry, suggests that nuclear fuel cycle facilities will have no difficulty in demonstrating compliance with risk limits currently proposed by ICRP.

iii. It is unlikely that underground uranium mining operations, however sophisticated, will be able to operate within the new ICRP recommended dose limits.

2.   The Draft ICRP Recommendations (February 1990)

2.1 Summary of the recommendations

Review of available studies of radiation effects on man suggest that the best estimate of the stochastic risk factor may be some four times higher than that used in ICRP 26 (1977). This is due to:-

-   Longer follow-up of the Japanese A-bomb survivors as the prime source of long-term stochastic (cancer) risk information.

-   A change to the use of the multiplicative model for projecting future risk of all cancers except leukemia.

-   the inclusion of secondary effects, eg hereditary stochastic effects in the primary risk coefficient.

-   reducing to an equivalent annual rate the sharply peaked mortality risk projected by the multiplicative model.

This re-analysis is accompanied by the suggested adoption of a Dose and Dose rate effectiveness factor (DDREF) which recognises the downward correction needed to produce risk factors applicable to exposures at low doses and low dose rates from data collected at high doses and high dose rates.

The recommendations also deal with non-stochastic effects, now called deterministic. There appears to be little change here from the 1977 recommendations with the LD_{50} being quoted as 3-5 Gray and the onset of deterministic effects at - 0.2 Sv (95% confidence).

For the first time, ICRP recommendations are also extended to include "potential exposure situations", ie accidents. In summary, they recommend a limit to fatality probability of $< 8 \times 10^{-4}$ per year for worker exposure and $< 5 \times 10^{-5}$ for public exposure. These are to be treated separately from occupational exposure risk and represent risks comparable to the long-term stochastic risks to workers and the general public from annual exposure at the recommended dose limit.

The recommended dose limits are given below (from page 5.16, Table 6)

TABLE 1 - NEW RECOMMENDED DOSE LIMITS[1]

Application	Dose Limit	
	Occupational	Public
# Effectance	* 100 mSv in 5 years 50 mSv in any 1 year	1 mSv per year, averaged over any 5 consecutive years
Annual dose equivalent in the lens of the eye the skin (100 cm^2) [2] the hands	150 mSv 500 mSv 500 mSv	15 mSv 50 mSv 50 mSv
Mean dose equivalent to the foetus [3]		5 mSv after diagnosis

NOTE 1. The limits apply to the sum of the doses from external exposure in the specified period and the 50-year committed dose (70-year for public exposure) from intakes in the same period.

NOTE 2. See paragraph 173 of the recommendations for details.

NOTE 3. This value relates to the occupational exposure of pregnant women, which is discussed in Section 5.3.3.

Since changed by the June Washington meeting to "Effective dose"
* Since changed by the June Washington meeting to "20 mSv in a year with some provision to allow year-to-year flexibility".

2.2 Comments on the Recommendations

The following points are made:-

-   The recommendations admit "a good degree of scientific judgement" is necessary in order to arrive at stochastic risk factors. Although they make the point that recommendations are meant to be neither optimistic nor pessimistic, the Commission has used the Japanese data as the prime source; if data from the irradiated spondylitic patients are used for the analysis the stochastic risk factors would be reduced by about a factor two.

-   As always, the ICRP review has been done comprehensively, but perhaps the real difficulty of interpretation is shown in the use of the Dose and Dose Rate Effectiveness factor (DDREF). There seems little doubt that correction is needed in making realistic estimates for risks at low doses and dose rates from data obtained at high doses. The evidence in the recommendations to justify the chosen factor of two is sparse. The conclusions drawn from it are very important. This invites the suggestion that excessive pessimism has been introduced.

- Despite the above, there is little to suggest that the ICRP recommendations should not be taken at face value, and this paper deals with the application to fuel cycle plant on this basis.

- The estimate of deterministic risk does not appear to have changed significantly and comment appears unnecessary.

- For potential exposure situations the Commission advocates the use of a "attribute vector". It introduces a variety of statistical deductions associated with the predicted mortality risk. It is clear that it is useful to have more than a single statistic to describe such a skew distribution as is produced by the multiplicative model, but it should not be forgotten that all criteria are only guides to judgement. The use of too many criteria readily lead to confusion in the minds of the expert and public alike.

- The key point is to recognise that the risk has to be estimated in terms of probability of occurrence x <u>consequence</u> (not dose). The usual consequence in work of this type is probability of death. It can be argued that this is emotive, fails to take cognisance of significant other effects like serious injury, and can lead to difficulties with the statistics of small numbers etc. Probability of death is used in probabilistic risk assessment (PRA) because the data available are unequivocal, not depending for example on medical opinion, and also because historical reviews have always suggested that if the fatal accident rate (FAR) is acceptable, then the rate of serious injuries, minor incidents etc will also be acceptable. This is because all the statistics are indicative of the safety culture that prevails.

- If consequence rather than dose is employed in the estimate, then there is no inconsistency between the two components of the criteria. A percentage reduction in consequence is exactly equivalent to the same percentage reduction in probability of occurrence, and the criticism made by the Commission on these grounds is invalid.

- There appears to be a lack of understanding revealed in the discussion about situations with only two possible outcomes, for example "death" or "survival". This situation can easily be handled using the argument presented in the previous paragraph.

It is also necessary to take care with risk criteria to avoid confusion between real risk and risk perception/aversion.

- It is noted that the use of stochastic factors in support of normal operations is not going to translate into supporting evidence for safe industry statistics if the consequences of occupational exposure <u>are regarded as accidents.</u>

3. **Implications for Design**

3.1 **Normal Operations**

In the UK fuel cycle facilities the relevant main plant design targets are:-

- Maximum whole body dose EDE+CEDE   15 mSv per year

- Maximum extremity dose   "   "   300 mSv per year

- Other organ doses within statutory limits

- Average exposure of workforce in plant less than 5 mSv per year.

These targets were introduced in 1989, but the major features have remained unchanged for more than 10 years. New plant designs have been executed to a significantly lower dose target than was demanded by ICRP 26, and therefore some reduction in revised recommendations has been anticipated for many years. Note that the above figures include maintenance activities.

BNFL has designed and constructed a wide variety of active plants over the last 10 years or so (over £2000 million at Sellafield alone) and so far there is no difficulty in achieving our design targets.

In France, the situation is similar for the reprocessing plant at La Hague. UP_3 and $UP_2$800 were designed so that during normal operation no employee receives a dose of more than 5 mSv per year.

3.2 **Accident Situations**

ICRP is now suggesting that as a result of accidents, a similar level of risk should be set to that implied for occupational exposure at the recommended limits. This comes to $< 8 \times 10^{-4}$ per year chance of death for workers and $< 5 \times 10^{-5}$ for the public. Viewed from a general industry stand point these are not stringent standards.

In the UK, BNFL has made use of accident risk criteria for the design and operation of plants for some years. These are intended to ensure the achievement of a FAR of 10^{-5} per year for workers. This rate is commensurate with the FAR achieved in the safest parts of industry. The following figures (taken from the UK Health and Safety Executive Paper "The Tolerability of Risk from Nuclear Power Stations") illustrate this.

TABLE 2:   SOME RISKS OF DEATH EXPRESSED AS ANNUAL EXPERIENCES
(UK DATA, 1987)

	per year	
Death by industrial accident to employees		
Deep Sea fishermen on vessels registered in the UK	8.8×10^{-4}	GB 1984
Quarries	3.9×10^{-4}	GB 1985
Coal extraction and manufacture	1.06×10^{-4}	GB 1986-7
Construction	9.2×10^{-5}	GB 1986-7
Agriculture	8.7×10^{-5}	GB 1986-7
All manufacturing industry including	2.3×10^{-5}	GB 1986-7
Metal manufacture	9.4×10^{-5}	GB 1986-7
Instrument engineering industry	2×10^{-6}	GB One death only in five years to April 1987

The French view is that ICRP recommendations should not encroach upon risk because this is a province of nuclear safety. In France, nuclear safety regulation is completely separate from that of radiological protection and this might lead to difficulties or confusion in regulatory function. The current view is that it is not realistic to base the safety of industrial installations on risk limits expressed in terms of probability of deaths.

However, the UK has many years experience in the chemical and other industries of working with risk criteria and in general it has not led to any particular difficulties. The key to success in this area lies in the recognition of the different roles of deterministic and probabilistic criteria, with the design and the safety of plant being essentially related to deterministic considerations. A second important feature is that the risk estimate has to be based on consequence and probabilities of occurrence which can be defended as pessimistic. This is not necessarily the worst possible case nor is it the expected or most probable outcome. With these caveats experience in the UK has shown there are no difficulties with working with risk criteria and none is anticipated with risk limits as conservative as those proposed.

4. Implications for Operation

4.1 UK Operations

These can best be judged by looking at historical information from operating plants. Recent BNFL experience in the UK is summarised in the table given below, which is taken from the most recent available published figures.

TABLE 3 - Summary of whole body exposure data for BNFL employees. The tabulated figures are the percentages of the workforce receiving doses in the stated range.

Whole body dose	Reprocessing		Reactors		Fuel Fabrication		Enrichment		Average of all exposure situations	
	1987	1988	1987	1988	1987	1988	1987	1988	1987	1988
0-5 mSv	66.4	66.9	53.3	56.8	82.8	86.8	100	100	74.9	75.7
5-15 mSv	24.3	27.0	36.8	40.5	14.9	12.8	NIL	NIL	19.0	20.5
15-30 mSv	8.9	6.2	9.9	4.7	2.2	0.4	NIL	NIL	5.9	3.8
> 30 mSv	0.4	<0.1	NIL	NIL	<0.1	NIL	NIL	NIL	0.2	<0.1

Please note that the higher doses are grouped in the way presented because BNFL in 1988 operated to a worker dose limit of 30 mSv. Recent decisions have led to this being reduced to 20 mSv with the responsibility to demonstrate that all worker doses are As Low As Reasonably Practicable over-riding, in anticipation of reductions in recommended dose rates. Data for operation and maintenance of BNFL's Magnox reactors are included for completeness.

With the possible exception of certain of the old plants in the reprocessing complex at Sellafield, it is anticipated that BNFL could operate to the proposed limits. The older plants will require significant operational control in order to meet the new proposals.

4.2 French Operations

Over the last six years the average external exposure of all of the workforce (Cogema + external contractors) in French reprocessing plants has largely remained below 1/10 of the current limit, ie 5 mSv/year. The average doses in the workshops or factories where the exposure level is highest are below 20 mSv/year.

The number of people with doses in excess of 20 mSv/year has been very small, varying between 0 and 0.6% of the monitored workforce with an average of 0.3% at the La Hague plant.

There will therefore not be any problem in conforming with the new limit so long as the working conditions of a very small number of people are modified.

In French reprocessing plants the level of internal exposure is extremely low and results almost exclusively from incidents. This is due to the efficiency of the containment systems and the systematic use of breathing apparatus in zones and during operations where there is a risk of dissemination of radioactive substances.

Nevertheless, a reduction in the annual limits of intake would involve profound modifications to the current system of individual monitoring for internal exposure by plutonium. Furthermore, the increase in complexity and cost of the surveillance system which would be required to guarantee observance of the annual limit would in no way serve to reduce the doses received.

In fuel fabrication plants in France where insoluble uranium compounds, such as UO_2 and U_3O_8 oxides are handled in the form of powders with a large number of operations being carried out manually, it is already difficult in certain workshops to observe the existing limits derived from atmospheric concentrations (LDAC).

These limits had already been divided by 8 in the national regulations on the basis of the values recommended in ICRP Publication No. 30.

A new reduction in the LDAC for insoluble uranium compounds would involve changes in technology and significant extra investment for the containment of powders and ventilation of working areas. The problems associated with emergency intervention and maintenance would also result in a large increase in operating costs in fabrication plants for mixed oxide fuels.

4.3 Uranium Mines

Although the problem of uranium mines is not dealt with in the new ICRP recommendations it is impossible to disassociate it from other installations of the fuel cycle. French national regulations, based on previous ICRP recommendations, set the total exposure (external exposure, radon and mineral dusts) on the basis of a limit of 50 mSv/year. Despite considerable efforts made in the mines with regard to ventilation and air purification, it should be noted that during the period 1984-88, 410 exposed miners out of 1276 monitored, that is 32%, exceeded the cumulative value of 100 mSv. Even supposing that the limit of 20 mj/year for potential alpha energy from radon and its daughter products remains unchanged, reducing the other two components contributing to exposure by a factor of 2 to 2.5 would have dramatic consequences on the exploitation of uranium minerals, particularly in underground mines. Because of the current technology, the mineral content and the low cost of uranium it is not possible to improve the current radiological situation in the French and African mines, in particular those in Niger.

The problem is likely to be the same in underground mines throughout the world. Exceptions are:-

- Canada, where total mechanisation is necessary because of the richness of the mineral.

- Australia, where extraction takes place in open cast mines.

- South Africa where uranium is extracted from the tailings from the diamond mines.

There would therefore be a upheaval in the geography of the world economy of exploitable uranium.

DISCUSSIONS

H. KOUTS, United States

I noted that this paper and the paper by Dr. Fay have a lot of similarities. I would therefore like to open the discussion on both these papers.

R. CUNNINGHAM, United States

If ICRP recommendations on lower occupational dose limits are adopted, the French believe, according to the paper by Dr. Berry and Dr. Henry, that there would be substantial problems in fuel cycle plants when uranium internal dose is involved. Similar problems would be experienced with plutonium resulting from operational events.

The summary of the U.K. situation, where no problem is anticipated, seemed to treat external dose only. Was the question of internal dose included in the assessment?

R.J. BERRY, United Kingdom

Yes the assessment included external dose plus 50 years committed effective dose equivalent. Our only problems in the U.K. come from old fuel handling plant where the standards of the past were not those which we would expect today. All our new plants have been designed and are operating to standards more stringent than the ICRP recommendations.

A. GONZALEZ, IAEA

In your verbal presentation you made two points that merit clarification:

1) An implicit suggestion that the dose reduction factor of two that would be recommended by ICRP is conservative on the basis of the UNSCEAR span of values.

However, it should be indicated that (i) the UNSCEAR values are derived from data at different dose levels and that, therefore, it is not surprising that the factors are very different, and (ii) the distribution of such values matches very well the value of two.

2) A suggestion that the individual risk limits recommended by ICRP are not stringent enough when compared with average rates at various industries.

This is an unfair comparison for two reasons:

a) You cannot compare a limit with an actual average.

b) Risk limitation is just one component of the ICRP system.

R.J. BERRY, United Kingdom

I do believe that the dose-reduction factor of two is a conservative figure. The experimental evidence covering a range of systems, human and non-human, has a range of values that extends well beyond ten. There is no good human evidence for any specific figure, apart from a feeling that it has to be greater than one because of the fact that we know that in every biological system we have seen some recovery during protracted radiation exposure for sparsely ionising radiation. I think it is a figure at the pessimistic end of possible numbers we could have used.

Then to the second comment you made. Even if you scale it down with a factor of ten, it would give us some difficulties but not a great deal of difficulties, so the argument would still be valid.

G. SILINI, Italy

I am referring to your statement that if radiation risk coefficients had been derived from the ankylosing spondylitis series, they would have been lower than those obtained from the Japanese series. As you well know, there are profound differences in the exposure regime between the two series. Precisely for these reasons ICRP felt the need to apply a reduction factor of two on the Japanese data. I cannot, therefore, accept your argument that seems to be a circular one.

R.J. BERRY, United Kingdom

I accept your basic premise, but even protraction of therapeutic radiation dose over a few weeks is not fully equivalent to protracted occupational exposure over many years.

D. BENINSON, ICRP

The range of "dose-dose rate-reduction factors" extracted from the relevant human data only shows a distribution around a value of 2, with a very low frequency tail going to higher values (the distribution is probably log-normal). The value of two does not correspond to the lower end, its a quite good average; there is a substantial number of data at approximately one.

F.J. TURVEY, Ireland

Can Dr. Berry remind us whether the health studies of workers at BNFL also included retired workers.

R.J. BERRY, United Kingdom

Yes. They include all workers employed at Sellafield before 1976.

H. KOUTS, United States

A question related to the mining industry and its ability to meet the more stringent recommendations. To what extent would solution mining be relied on to solve this problem. That is usually referred to as one possible solution.

R.M. DUNCAN, Canada

Regarding solution uranium mining, limited experience in Canada has shown that this type of mining of low-grade ore bodies actually resulted, on average, in higher exposures and doses than conventional mining techniques. The main reason was the need to handle large quantities of liquids in confined areas.

R.H. CLARKE, United Kingdom

I want to emphasize that although uranium mines present a difficult problem for protection, it is an area from which we have direct epidemiological evidence of harm. Although you say that historical levels of radon daughters were higher, we have a linear dose-response relationship for lung cancer against WLMs, including data at levels corresponding to current exposures.

So I say that the risks are more real for uranium and non-coal miners than for other nuclear workers.

D. BENINSON, ICRP

I agree that underground mines are a real problem. In some cases even external radiation are a substantial contribution making reductions very difficult.

R.M. DUNCAN, Canada

It is the exposures in uranium mines of 15 to 20 years ago which have resulted in recent deaths of uranium miners, not current levels of exposure.

INTERFACE BETWEEN RADIATION PROTECTION AND NUCLEAR SAFETY

Gunnar Bengtsson
Swedish Radiation Protection Institute
Stockholm, Sweden

Lars Högberg
Swedish Nuclear Power Inspectorate
Stockholm, Sweden

ABSTRACT

Interface issues concern the character and management of overlaps between radiation protection and nuclear safety in nuclear power plants. Typical examples include the selection of inspection and maintenance volumes in order to balance occupational radiation doses versus the safety status of the plant, and the intentional release to the environment in the course of an accident in order to secure better plant control. The paper discusses whether it is desirable and possible to employ a consistent management of interface issues with trade-offs between nuclear safety and radiation protection. Illustrative examples are quoted from a major Nordic research programme on risk analysis and safety rationale. These concern for instance in-service inspections, modifications of plant systems and constructions after the plant has been taken into operation, and studies on the limitations of probabilistic safety assessment. They indicate that in general there are no simple rules for such trade-offs.

INTERFACE ENTRE RADIOPROTECTION ET SÛRETE NUCLEAIRE

RESUME

Les questions d'interface concernent le caractère et la gestion des domaines de chevauchement entre la radioprotection et la sûreté nucléaire dans les centrales nucléaires. On peut citer notamment à cet égard le choix de l'importance à donner aux activités d'inspection et de maintenance afin de mettre en balance les doses de radioexposition professionnelle et l'état de sûreté de l'installation, et le rejet délibéré dans l'environnement au cours d'un accident afin de garantir une meilleure maîtrise de l'installation. Les auteurs examinent s'il est souhaitable et possible de recourir à une gestion cohérente des questions d'interface avec des arbitrages entre la sûreté nucléaire et la radioprotection. Des exemples caractéristiques sont empruntés à un important programme de recherche nordique sur l'analyse des risques et les principes de base de la sûreté. Ils portent, par exemple, sur les inspections en service, sur les modifications des systèmes et structures de l'installation après sa mise en service et sur les études relatives aux limitations de l'évaluation probabiliste de la sûreté. Ces exemples montrent qu'il n'existe pas, en général, de règles simples pour de tels arbitrages.

INTRODUCTION

Radiation protection and nuclear safetyhave considerable interfaces. Both areas refer to the prevention or mitigation of detrimental effects by radiation on human beings. Nuclear safety, however, has more emphasis on prevention and refers exclusively to nuclear installations. Its overriding objective is to protect individuals, society and the environment by establishing and maintaining an effective defence against radiological hazards [1]. A multi-level defence-in-depth is achieved through the proper design, construction, operation and maintenance of nuclear installations.

For decisions on nuclear issues, it is often of interest to consider the character of the interface between nuclear safety and radiation protection, and the management of the pertinent issues. Typical interface issues include the following:

* can the volume of inspections and maintenance be reduced in order to reduce occupational doses even if there is an associated safety impairment?

* is it unwise to introduce backfitting of safety features because the backfitting operation causes high occupational doses?

* can an intentional radioactive release to the environment in the course of an accident be justified because it enables a safer management of the accident?

Interface issues have been extensively discussed in the literature [2]. Attempts at developing coherent safety rules [1] similar to the recommendations by the International Commission on Radiological Protection, ICRP, and discussions on formal trade-offs between radiation protection and nuclear safety [3] have spurred new interest in the interface issues. In this paper we discuss some basic questions:

* Do political decision makers use an integrated and consistent management of risk associated with radiation protection and nuclear safety?

* Do decision makers within nuclear power plants manage risks consistently, with discernible rules to account for radiation protection and nuclear safety issues?

* Are the existing data on risks, cost and other factors reliable enough to permit a systematic and meaningful treatment of issues on trade-off between radiation protection and nuclear safety?

In the following we claim that the answer to all three questions is NO. There are no comprehensive and widely accepted models, based on calculations of trade-offs between risks and benefits, despite the long prevalence of the ICRP principles of optimisation. To a large extent we support this judgment by data from a major Nordic research program [4].

DO POLITICAL DECISION MAKERS USE A CONSISTENT INTERFACE MANAGEMENT?

Many aspects enter into political decisions concerning nuclear power, such as

*   risks to human health and other environmental risks
*   the degree of certainty to which these aspects are known
*   public perception of the risks
*   costs of the various prevention or mitigation measures
*   legal considerations.

The decision maker must account for all relevant aspects despite their being incomparable in any objective sense. In a naive perception of political realities, it might be expected that the decision maker reviews all relevant aspects and seeks a systematic trade-off. In the real world, the decision maker does not have the time nor the money for the analysis, and is meeting so many difficulties that she or he recurs to gross simplifications [5]. The whole science of policy analysis [6] has sprung out of the need to aid the real political decision making as much as possible. Several case studies including radiation protection and nuclear safety [7] support the idea that the systematic trade-off is not a political reality.

This is not surprising. Even the limited issue of "risks to human health and other environmental risks" in nuclear power interface decisions contains many factors that are quite different in character. In radiation protection, there is a strong concentration of the interest to low radiation doses which are mainly associated with late health effects such as cancer and hereditary disease. In contrast, the political interest related to nuclear accidents is focussed on acute radiation effects and long-term extensive ground contamination. The latter is evidenced by the Swedish government decision from February 1986 where these effects were explicitly mentioned as the basis for measures to restrict large releases in the case of nuclear accidents. There is no simple way of trading off late health effects versus ground contamination and acute health effects. Indeed, the valuation of such effects does not appear to be linear. The social disruption caused by large-scale, long-term evacuation of contaminated areas can hardly be expected to be quantified in a way that meets wide political consensus. Recent decisions in the Netherlands [8] and Denmark [9] support a special political weighting of large accidents. They demand that the probability of any large accident be kept below the inverse square of the number of ensuing serious injuries or fatalities, apart from a proportionality constant.

A corner-stone in radiation protection management is the requirement to keep radiation doses as low as reasonably achievable. This is also called to optimise radiation protection, and cost-benefit analysis is a permitted tool. There are, however, many constraints to the optimisation and it cannot easily be extended to the field of nuclear safety. At least in Finland and Sweden an equivalent optimisation is not allowed. The political decision makers have set strict boundary conditions for it. A lowering of the level of safety is not permitted even if it would

be the result of a cost-effective trade-off against e.g. the energy
availability of the reactor. The alternatives that remain in order
to attain greater cost-effectiveness in plant operation are either
to lower the cost at a constant safety level or to increase the
safety at a constant cost.

There are thus abundant examples that the political
decision makers reserve the right to decide whether any trade-off
should be permitted between radiation protection and nuclear
safety. This leaves little room for trade-offs performed by
authorities and plant managers.

In an earlier study [10], we have discussed several cases
where optimisation of a simple type might have been applied, but
in fact did not have a decisive influence on the prevention of
accidents and mitigation of their consequences. We proceed now to
some cases of operational radiation protection and probabilistic
safety analysis.

DO PLANT MANAGERS APPLY OPTIMISATION IN RADIATION PROTECTION?

If the trade-off between various factors is so difficult
for politicians, could it be easier in the simpler case of plant
management? To study this, a first step may be to look at the
simple case of radiation protection decisions. If not even these
are amenable to simple trade-offs, it seems very far off that that
a systematic trade-off could be extended to include also nuclear
safety. Indeed, a Nordic study [11] in Finland and Sweden does not
support that formal optimisation is used in radiation protection
practice. In the following, the main results are presented. They
concern applications in:

*   the use of protective clothing and equipment. This involves
    only radiation protection and is included here to illustrate
    an optimisation method, where there is no confounding with
    nuclear safety.

*   in-service inspections, which involve both radiation
    protection and nuclear safety

*   modifications of plant systems and constructions, which also
    involve both radiation protection and nuclear safety.

Protective clothing and equipment

Information from the Nordic nuclear power plants was obtained
and presented concerning the use of i.a. boundaries for contaminated
areas, protective clothing, respirators, and temporary shielding.

In addition a cost-benefit formalism was developed as a
case-by case decision aid for the use of temporary protective
equipment. An example of the result of its use is given in figure
1, which shows the cost effectiveness of lead blankets with a dose
reduction of 33 %. The cost of the dose reduction per unit dose
equivalent is given as a funtion of the initial dose equivalent
rate, with the time the blanket is used as a parameter. In Nordic
radiation protection, it has been considered appropriate that the

radiation protection authorities advocate an expense of at least 20 000 USD per mansievert averted, or about 100 000 FIM. If this is applied, it is e. g. advisable to use lead blankets with a dose reduction of 33 % for a work operation of 1 hour if the initial dose rate exceeds 1.3 mSv per hour.

The results show that the protective measures applied today at Nordic nuclear power plants are fairly compatible with the internationally recommended principles and guidelines for optimisation. It is not clear whether this is fortuitous or a result of radiation protection judgment implicitly influenced by international optimisation principles.

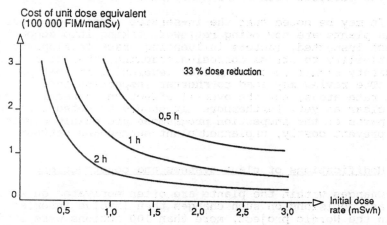

Figure 1. The cost per unit dose equivalent averted when one lead blanket is used and its dependence of dose rate and working time. The total costs caused by the use of the blanket (including the personnel costs of installation) are 40 FIM (1 FIM is about 0.2 USD) for each time it is used. The dose reduction achieved through lead blanket use is 33 % of the dose received without the blanket.

In-service inspections

In-service inspections are considered to be of great importance to plant safety and reliability. In the optimisation of their radiation protection, questions of nuclear safety are almost always involved and interface issues thus arise. In the Nordic study, information was collected on regulations and practices for in-service inspections, radiation dose statistics, radiation protection measures, and decision criteria for the optimisation of radiation protection at the Nordic power plants.

One of the findings was that exposure of workers during in-service inspections contributed about 15 % to the annual collective dose. The individual doses are sometimes high, and protective actions must be continually considered, particularly for the insulation personnel and for inspection personnel who work at several different plants during maintenance periods.

Optimisation in the strictly formalized sense had not been used at the plants. Instead, practical optimisation based on operating experience was applied. Besides dosimetry and other direct radiation protection measures, one of the most important actions for controlling doses during in-service inspections seemed to be an active work management programme.

Automation of some parts of the inspections is considered a step forward from the radiation protection point of view. Reconsideration of the frequency and extent of the traditional inspection program would mean additional progress bu involves difficult interface decisions. Such decisions might be facilitated if data were collected and evaluated systematically, e.g. concerning the radiation doses and the costs of the protection alternatives possible with the new methods.

It may be noted that the in-service inspection programmes at Swedish plants are now being reviewed, taking into account, for each object inspected, factors influencing crack development (such as susceptibility to stress corrosion cracking, thermal fatigue, etc) and the safety significance of a leak developing in that particular position. The review may lead to shorter inspection intervals in some high dose rate areas, but the overall effect on inspection doses is far from clear as yet. Furthermore, it should be noted that substanstantial parts of the inspection programmes are mainly carried out in order to prevent costly, unplanned plant shut-downs between refueling outages.

Modifications of plant systems and constructions

Changes within the plants are often motivated on safety grounds, and decisions on such changes then involve interface issues. In the Nordic project, more than 100 actions were studied, ranging from small modifications made just to improve radiation protection, to larger new constructions. The resulting view was that actions to reduce doses were based on more direct needs than optimisation considerations. Such needs could concern e.g. high local or general dose rates, or operational or safety related factors. Optimisation was hardly ever done quantitatively. Instead, it had been more like an intuitive process, based on the experience and skill of the radiation protection staff.

Actions were generally considered to be cost-effective by the individuals involved in the decisions. The study revealed that some of the actions involve rather high costs, reflecting the relatively small weight given by the plant operators to such input to their decisions as the relatively small collective radiation doses involved.

Conclusions

The optimisation studies showed that formal optimisation is not extensively used in practice in the Nordic countries. One reason is that the procedures are too complex, and this might be alleviated by better methods, maybe computer based. Another important factor is that the radiation detriment is given little weight in the decisions and its relation to cost not recognised or assessed in detail. This means that considerations on the interface between radiation protection and nuclear safety have not presented any major difficulties to the plant decision makers.

ARE RISK DATA RELIABLE ENOUGH TO PERMIT INTERFACE TRADE-OFFS?

A formalistic treatment of the trade-off between radiation protection and nuclear safety requires that quantitative estimates of cost, risks and other factors are available in the two fields. Can such estimates be made for the field of nuclear safety?

Systematic probabilistic safety analysis, PSA, provides estimates of the frequencies of events leading to core damage. Higher level analysis may also provide estimates of public risks. However, the uncertainties in such estimates typically span over many orders of magnitude. This is demonstrated in a recent Nordic study on Finnish and Swedish nuclear power plants [12] as well as in the USNRC NUREG-1150 study and subsequent reviews [14, 15]. There is no doubt that the state-of-the-art in PSA has reached a certain degree of maturity, and PSAs are practically used for e. g. planning and reviewing of plant modifications. At the same time there are many remaining problems and limitations in using probabilistic techniques. Some of them are intrinsic and difficult or impossible to overcome, while other are matters of practice and thus bound to be resolved as understanding of analytical methods becomes more widespread.

Examples of the intrinsic and practical limitations are given below. Sometimes both types of limitation may affect an aspect of a PSA. The compilation is i. a. based on retrospective reviews of Swedish, German, British and United States PSAs, made within the framework of the Nordic study. The reviews showed, for instance, that only one out of six Swedish PSAs contained a thorough analysis of human interactions. Other benchmark studies and sensitivity analyses showed that the modelled core damage frequency varied considerably between modelling teams. For instance, the range between the estimates including 90 % confidence intervals covered 3 decades for the simultaneous failure of redundant motor operated valves (figure 2), and 5 decades for the most dominant accident sequence at a Swedish plant. The latter was, however, selected because the uncertainties were expected to be very large.

Intrinsic limitations

Incompleteness. Incompleteness of PSA originates from the technical complexity of large systems, from difficulties to identify, model and quantify all potentially significant internal interactions between systems, components, human beings and corresponding external interactions with environment, and from possible errors in the use and review of PSA.

Obviously, there are no guarantees that all significant accident sequences can be identified, but credibility of the studies is expected to increase with time. Integration of new operating experience, i.e. calibration based on the real world, within the concept of living PSA will contribute to reduce incompleteness uncertainties. Sensitivity analysis may also be employed to some extent in this context.

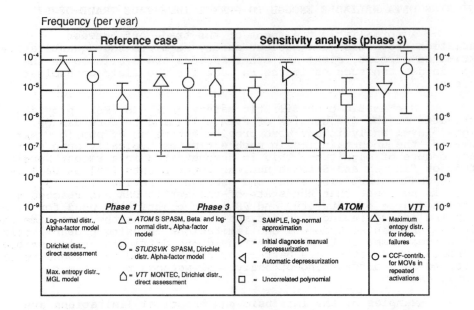

Figure 2. Results by three Nordic modelling teams (ABB ATOM, STUDSVIK, and VTT) from uncertainty and sensitivity analysis for the most dominant sequence from the Forsmark 3 PSA. The sequence is dominated by common cause failure contributions for motoroperated valves and by operator failure to initiate manual depressurization. The selected sequence is not typical for the Swedish PSAs and the associated uncertainties are expected to be very large. In Phase 1 of the study, emphasis was on the generation of uncertainty distributions, and methods and data quantifying common cause failures and human interaction errors were selected quite freely by the participating groups of analysts. In Phase 3, the participants used models, data and computer codes considered optimal. All groups used Monte Carlo simulation codes for the propagation of uncertainty distributions. The estimated frequencies differed substantially in phase 1 but were more consistent in phase 3 as a result of modified approaches. From a practical point of view, replacement of the existing manual depressurisation by an automatic one would reduce the estimated frequency by a factor of 20.

Data Base. The data bases will always be incomplete. This applies also to the information on relatively frequent failures, at least with respect to detailed knowledge. Furthermore, estimates of probabilities of radioactive releases of different magnitudes in case of severe core damage accidents have to rely extensively on expert judgements. The uncertainties associated with such judgements is very large in many cases [14, 15].

Human Interactions. The influence of human errors can be modeled and quantified only to a limited extent in PSA. This is particularly true for errors of commission. This class of errors include for example serious errors in the decision-making in the control room with detrimental effects on plant safety state or accident progression. Such errors of commision may contribute significantly to overall risk [15]. Also, it is difficult to model the human ability to act innovatively, to solve problems and to correct mistakes.

Common Cause Failures - a subset of dependencies. The common cause failures are very important in the case of redundant safety systems or components - a common cause failure may mean that all systems or redundant trains within a system are becoming inoperable at the same time, and the redundancy is thus lost. A variety of different causes may be hidden behind each common cause failure contribution, and the analysis of common cause failures is quite complex.

Similarly to human failures, common cause failures may result from an infinite spectrum of possible scenarios based on hypothetical root causes and coupling mechanisms. Although there are models and some empirical data that can be used to estimate the contribution of common cause failures, this type of failure still represents one of the major uncertainty sources in quantitative risk estimates based on PSA.

Practical limitations.

Consistency. Experience shows both similarities and differences between the PSAs performed. Efforts have recently been made in Sweden to separate those discrepancies which reflect differences in design and operation from those which are due to differences in modelling approaches. The study showed that better consistency can be achieved.

Realism. The objective of a PSA is to provide a plant-specific realistic model of accident propagation. This means that unduly pessimistic assumptions should be avoided.

In reality the uncertainties are frequently handled by means of safety margins, the magnitude of which is not known. In addition, proper credit is not always taken for normally operating systems which can prevent or mitigate the accident. The final quantitative result might then be distorted.

Uncertainty. Numerous examples can be given of careless treatment of uncertainties. Clear distinctions should be made between the parametric, modelling and incompleteness uncertainties and their relative importance. This is often not the case. The analysts are not always aware of which type of framework (frequentist or subjectivist) they actually apply and this may lead to inconsequences. The NUREG-1150 study [14] demonstrates the effort necessary to address uncertainties in a systematic way.

Human Interactions. Apart from having intrinsic limitations, modelling of human interactions is subject to shortcomings which can be improved. This applies in particular to qualitative modelling of operator tasks in the control room, where depending on the particular situation different types of behaviour (skill-, rule-and knowledge-based) may be expected.

There is frequently a pessimistic or lacking treatment of recoveries in the PSAs, that is, of the restoration and successful operation of equipment which was initially unavailable. The available data for human interactions apply only to well structured tasks.

Dependencies. Treatment of dependencies is not always well structured and consequent. Some common cause failure models are based on questionable assumptions or are being applied without proper regard to their limitations. Frequently, methods for quantification of common cause failure contributions are not compatible with the amount and quality of information available from data sources.

External Events. The lack of relevant data and the complexity of the problem create large uncertainties. The treatment of seismic hazards in low seismicity regions appears to be one of the most difficult. These hazards may be among thed dominant risk contributors. Progress in this area is expected due to increasing knowledge about seismicity, fracture mechanics and seismic response.

Time Dependencies. The basic logical models of PSAs (event trees, fault trees) can only to a limited extent simulate the numerous types of time dependencies which are involved and which may be important. Supplementary analyses of these aspects are seldom made within the PSAs.

Validity and applicability of risk estimates

Current research projects carried out in Nordic countries and elsewhere clearly demonstrate the spectrum of problems encountered when comparing different PSAs. Thus, directuse of plant-specific numerical results in the absolute sense should be made with great caution, having in mind a wide spectrum of intrinsic and practical limitations as well as the involvement of subjective judgement in almost all tasks of a PSA.

There are however many good examples of design and procedural deficiencies at nuclear power plants which have been disclosed and remedied as a result of PSA work. Modifications are often made at the plants in order to achieve a more balanced risk profile. Such use of PSA techniques has indisputable advantages. In this context the use of PSA results for decision-making is based on relative criteria. These are, as opposed to absolute criteria, not totally dependent on the exactness of predictions and consequently less sensitive to uncertainties. Also, PSA methods have been used to improve the technical specifications which govern preventive maintenance and repairs in safety systems of Finnish and Swedish reactors [13].

Thus, the performance of in-depth, plant-specific PSAs makes sense as a basis for plant-specific risk management efforts. For example, issues about completeness should not prevent backfitting efforts to achieve estimated core damage frequencies below one in a hundred thousand per reactor operating year, as this is a necessary but not sufficient condition to achieve what has been recognized as a reasonable level of safety [1] with respect to the protection both of the public health and safety and of the investment in nuclear power generation. Furthermore, a good understanding of uncertainties identified in a PSA makes possible a more sophisticated approach to risk management, addressing not only major contributors to risk, taken as point values, but also con- tributors associated with large uncertainty bands. One approach to risk management in such cases would be that a small increment in the probability of an event with small or moderate consequences might be considered tolerable as a trade-off for a substantial reduction of a large uncertainty band associated with a high- consequence event, although this event has a low point value estimate of probability. The filtered containment venting systems installed in some countries exemplify such an approach to the containment overpressure failure uncertainty.

As a consequence of the intrinsic and practical limitations of PSA, considerable controversy surrounds the reliance on the bottom-line results of PSAs as a quantitative measure of overall risk and, consequently, the use of formal quantitative safety goals in the regulatory process: "One should resist one-digit statements about safety." The same arguments apply to quantitative trade-offs between radiation protection and nuclear safety.

Introduction of formal risk criteria for licensing is, in our opinion, hardly motivated in view of the problems outlined related to the inadequate precision of presently available tools to demonstrate formal compliance with safety goals - and also the difficulties associated with specifying such goals. Implementation of such goals would require detailed specification of analysis procedures, a formidable and practically impossible task, the realisation of which would not encourage future developments and hardly promote safety improvements.

Rather, the safety goals should be considered as one factor in arriving at regulatory judgement. This has a parallel in our opinion that trade-offs between nuclear safety and radiation protection should be based on qualitative more than quantitative considerations.

DISCUSSION AND CONCLUSIONS

The cases presented above concern several areas of strong interfaces between nuclear safety and radiation protection. We have earlier shown [10] for the case of optimisation in the prevention and mitigation of accidents that political, commercial and engineering judgments often lead to decisions far away from those suggested by simple optimisation rules. In this paper we broaden the issues considered and include more studies on in-service inspections and modifications of plant systems and constructions. Often, these have a strong component of interface between nuclear safety and radiation protection. We also illustrate the radiation protection optimisation methods concerning protective clothing and equipment. The results as well as recent results on the uncertainties in probabilistic safety analysis point in the same direction.

The general experience is that radiation protection and nuclear safety decisions concern many very different aspects, even if one considers only the risks to health and the environment. Trade-offs between these aspects are too complex to be guided by simple optimisation rules, and e. g. in Sweden and Finland have been partly forbidden by the politicians. Even at the nuclear power plant level, simple optimisation rules are not used in interface issues. One good reason why this is not done lies in the evaluation of low probabilities. Studies in probabilistic safety analysis indicate that calculated core melt frequencies may have uncertainty bands spanning several decades, and this effectively precludes quantitative trade-offs. The uncertainties become still larger when extending the PSA to estimates of radioactive releases and resulting doses to the public.

To soothe the defaitistic message somewhat, we would like to add that quantitative trade-offs might be meaningful in some cases. If the radiation consequences are of a similar character and the probabilities can be well estimated using established data bases, it might be rational to attempt systematic allocation of resources for nuclear safety versus radiation protection. We imagine that such cases might be at hand e. g. in the safety of small and well defined systems where the maximum consequences of any accident are fairly small and limited for physical reasons.

ACKNOWLEDGEMENTS

We have drawn very heavily upon material collected in a large Nordic research program [4] and are indebted to those involved in this program, in particular the project managers Olli Vilkamo and Stefan Hirschberg.

REFERENCES

[1] Basic safety principles for nuclear power plants. IAEA Safety Series No. 75 - INSAG-3. IAEA, Vienna 1988

[2] OECD Nuclear Energy Agency: Interface questions in nuclear health and .safety. Proceedings of an NEA seminar. OECD, Paris, 1985

[3] International Atomic Energy Agency: Extension of the principles of radiation protection to sources of potential exposure. IAEA Safety Series 104. IAEA, Vienna, 1990

[4] Bengtsson G (ed.): Risk analysis and safety rationale. Final report of a joint Nordic research program in nuclear safety. Nordic Council of Ministers Report NORD:91. Swedish National Institute of Radiation Protection, Box 60204, S-104 01 Stockholm, Sweden, 1989.

[5] Lindblom C E: The science of "muddling through". Public Administration Review 41 (1959), 517-26

[6] MacRae D and Wilde J: Policy analysis for public decisions. Lanham: University Press of America, 1985

[7]   Bengtsson G: Integration of economic and other aspects in
        decisions to regulate genotoxic substances. In Proceedings of
        Symposium on Management of risk from genotoxic substances in
        the environment, Stockholm 3-5 October, 1988. Swedish
        National Chemicals Inspectorate, P O Box 1384, S-17127 Solna,
        Sweden, 1989, 291-306

[8]   van Kuijen C J: Risk management in the Netherlands - a
        quantitative approach. In Proceedings of an IAEA/UNEP/WHO
        workshop on Assessing and managing health and environmental
        risks from energy and other complex industrial systems, Paris
        13-17 October, 1987. IAEA-TECDOC-453. IAEA, Vienna, 1988,
        149-162

[9]   Miljöstyrelsen: Kvantitative og kvalitative kriterier for
        riskikoaccept (Probability and qualitative criteria for risk
        acceptance) (in Danish). Miljöstyrelsen Miljöprojekt Nr. 112,
        Strandgade 29, Dk-1401, Köbenhavn K, 1989.

[10]   Bengtsson G and Högberg L: Status of achievements reached in
        applying optimisation of protection in the prevention and
        mitigation of accidents in nuclear facilities. In Proceedings
        of Ad hoc meeting on the Application of optimisation of
        protection in regulation and operational practice. OECD/NEA,
        Paris, 1988

[11]   Vilkamo O (ed.): Application of the optimisation principle to
        radiation protection at nuclear power plants. Nordic Council
        of Ministers Report. Strålsäkerhetscentralen, Postbox 268,
        SF-00101 Helsinki, 1990

[12]   Hirschberg S (ed.): Treatment of dependencies, human
        interactions and associated uncertainties. Summary report of
        the Nordic NKA project Risk Analysis. Nordic Council of
        Ministers Report. ABB Atom, Box 53, S-72163, Västerås, Sweden
        1990.

[13]   Laakso K (ed.): Optimisation of technical specifications
        using probabilistic methods - A Nordic perspective. Nordic
        Council of Ministers Report. Statens tekniska
        forskningscentral, SF-00101 Esbo, 1990

[14]   Severe accident risks: An assessment of five U.S. nuclear
        power plants. USNRC NUREG-1150, second draft. US Nuclear
        Regulatory Commission, Washington D.C. 1989.

[15]   Special committee review of the Nuclear Regulatory
        Commission's Severe Accident Risks Report (NUREG-1150).
        USNRC NUREG-1420, US Nuclear Regulatory Commission, Washington
        D.C. 1990.

DISCUSSIONS

H. KOUTS, United States

So we have an additional problem connected with optimisation thrown out by the authors of the last paper. What is the value of sophisticated decision making tools?

A GONZALEZ, IAEA

My question relates to the presentation by Dr. Bengtsson and Dr. Högberg.

Your paper presents two implicit conclusions:

a) There are no trade-offs in safety-related decisions but just "judgement", either political or engineering.

b) Even if a formalised trade-off should be required, either the unavailability of data or the uncertainties of the assessments would make such trade-off impossible.

I question these conclusions on the following grounds:

a) The trade-off always exist: it can be explicit, when decision are formalized technically, or implicit, when the decisions are of a dogmatic nature. Formal decisions on technical ground have the advantage of logical structure, traceability and rationality. Technical professionals should adhere to this principles.

b) Unavailability of data and uncertainties thereof relate to the technical knowledge of a problem and not to the decision-making process for solving the problem. Whether we use a formal, technical decision-making process or a dogmatic one is irrelevant for the existence or data and uncertainties and vice versa. But, of course, if a formal technical decision-making process is used, such lack of knowledge become apparent, while dogmatic decision-making conceals it.

L.G. HÖGBERG, Sweden

1) Of course, trade-offs are made in the type of decisions discussed in our paper; our point was that such trade-offs are not based on numerical optimisation models.

2) We do indeed support that uncertainties should be explicitly described in the material used as a basis for regulatory decisions. In fact, we point out in our paper that explicit descriptions of uncertainties enable a more sophisticated approach to risk management.

3) We support a rational approach to risk management and optimisation per se - but we emphasise in our paper two fundamental weaknesses of numerical decision-making models:

a) The scientific uncertainties in PSA estimates are still too large to allow meaningful, quantitative trade-off calculations.

b) There are too many social and economic factors involved, widely different in character, to allow meaningful development of practically acceptable numerical decision-making models.

L.G. BENGTSSON, Sweden

Dr. Gonzalez also asked if we are satisfied with the present state of affairs. My answer is yes. What we need is not more scientists trying to deal with political matters, making multiattribute utility analysis, etc. What we need is more communications. This belongs to the political arena. In this process we, the scientific community should provide the data, the uncertainties involved, etc.

P.Y. TANGUY, France

Je partage le point de vue de M. Bengtsson : la situation actuelle, en ce qui concerne les processus d'arbitrage entre sûreté et radioprotection. Pour moi, cela vient de ce que, dans chaque cas concret, il y a l'un des deux aspects qui domine la décision : la sûreté pour l'inspection de composants de réacteurs, la radioprotection pour les rejets d'effluents. L'autre aspect est alors pris en compte comme une contrainte extérieure.

Je ne souhaite pas une formalisation, qui n'apporterai rien. Je ne suis pas opposé à un certain "flou", plutôt qu'à une explication complète des processus de décision, dans la mesure où des aspects sociaux et d'opinion publique sont en jeu.

Session 5

CONCLUSIONS AND RECOMMENDATIONS

Séance 5

CONCLUSIONS ET RECOMMANDATIONS

Chairman – Président

K.B. STADIE
(NEA)

PANEL DISCUSSION - TABLE RONDE

K.B. STADIE, NEA

Good afternoon, ladies and gentlemen. We would like to begin with the Panel. First of all, I should like to applaud the stamina of those in the audience who have borne with us through one and a half days of intense debate which we will now try to wrap up with this Panel discussion. The Panel is made up of the four session chairmen, and of four representatives, which we have carefully selected from among the different regions of the OECD and the different discipline interested in the question of interface in nuclear health and safety. We have debated, I think, over the last day and a half, all the major, and many of the minor issues which divide or unite radiation protection and nuclear safety and I hope that we will not go on to repeat these issues one by one. The main questions for us to discuss now: Where do we go from here? and how do we resolve the remaining issues? I personally do not know and I turn for inspiration to the Panel members and the audience. What I know is that we shall not, and we must not, wait another five years to have a debate of the kind which we have had for the past two days. If there is going to be a nuclear renaissance, it will certainly be necessary to resolve the remaining questions between us sooner than that. Otherwise, the nuclear renaissance will wait even longer than it will have to wait anyway.

As regards the organisation of our discussion, we will begin with summaries of the four chairmen and we will follow the sequence of the Programme of the Seminar and address first the question of what are the interface issues. We will then discuss the basic objectives and policies. After that, we will summarise achievements of radiation protection and nuclear safety objectives in the regulatory and technical practice. To each of the four summaries we have asked the other Panel members to comment on, to add something, if they feel the need to do so. At the end I hope that we will have sufficiently well defined the unresolved problem and made recommendations for further work.

I now turn to Monsieur Cogné to talk about the interface issues. Monsieur Cogné.

F. COGNE, France

Merci M. le Président, j'ai donc eu l'honneur de présider la première séance et nous avons eu trois remarquables exposés que je ne vais pas essayer de résumer, parce que je les trahirai, et puis surtout parce que j'ai un avantage important ici, qui me pousse à la paresse, c'est que les trois présentateurs de cette première séance sont à la Table ronde et qu'ils pourront beaucoup mieux que moi préciser si besoin est des points importants. Je veux dire Dr. Clarke, Dr. Kouts et le Profeseur Silini.

Je voudrais dire d'abord que cette séance a permis d'aborder au fond et dès le début le sujet même de ce Séminaire, c'est-à-dire, comme le disait le Président tout de suite, les interfaces entre sûreté et radioprotection. La présentation de MM. Clarke et de Birkhofer a montré que la sûreté nucléaire et la radioprotection ont évolué en parallèle. Même si les philosophies pour la gestion du risque sont différentes, et on l'a bien revu encore aujourd'hui, les interfaces sont nombreuses et il est important que les deux communautés arrivent au moins à bien se comprendre, car finalement elles sont attachées au même objectif qui est d'obtenir que l'utilisation de l'énergie nucléaire ne conduise pas à des risques significatifs pour l'homme et pour l'environnement. or la discussion – je m'attacherai uniquement à la discussion – a bien montré que ce n'était pas sur l'objectif qu'il peut y avoir des différentes appréciations mais sur les moyens d'y arriver. De cette discussion, je retiendrai trois points.

i) Il y a pour les gens de la sûreté une réelle ambiguïté quand ceux de la protection disent que l'optimisation est égale à ALARA. On l'a revu, on en a discuté tout le long, il en a été question encore aujourd'hui, pour moi c'est un point qui m'est apparu extrêmement clairement au cours de ce Séminaire et tout particulièrement dès hier matin.

L'exemple des rejets d'effluents, sur lequel nous sommes revenus aujourd'hui, des centrales nucléaires montrent qu'ils n'ont jamais fait l'objet d'une optimisation au sens mathématique du terme, mais seulement d'une minimisation, que les industriels ont accepté d'améliorer continuellement les méthodes de traitement en prenant au pied de la lettre le principe aussi bas que possible. Aucune optimisation n'aurait d'ailleurs, me semble-t-il, conduit à des valeurs aussi basses que 1 µSv par an par individu en Allemagne. Mais personne ne paraît s'être posée la question économique. En fait, comme l'a dit M. Vuorinen, le terme même d'optimisation n'est pas un bon terme, puisque ce n'est pas ce qu'on fait.

ii) Le deuxième point que je voudrais souligner de la discussion est que la réduction des doses aux travailleurs est beaucoup plus due à l'application par les concepteurs et les exploitants du principe ALARA que par la diminution des doses limites elles-mêmes. Et ceci a été redit plusieurs fois au cours de ce Séminaire, et je crois que c'est un thème qui méritera certainement une réflexion complémentaire.

iii) Enfin, ce qui rejoint un peu ce que je viens de dire pour les gens de la sûreté, on s'est demandé quelle peut bien être la signification de la dose collective pour de très faibles doses individuelles, ceci a été aussi un peu abordé et d'ailleurs dans le papier de M. Clarke, en filigrane, d'un côté on parle d'une coupure à 20 µSv par an, mais de l'autre on pousse à des valeurs plus basses.

Le Dr. Gonzalez nous a accusé d'être deux mafias qui utilisent le même dictionnaire, mais ne mettent pas les mêmes choses sous les mêmes termes. Je crois qu'en fait ce n'est qu'une question d'éclairage différent. La sûreté cherche à traduire en dispositions techniques les limitations de risque, alors que la radioprotection essaie de traduire en effets aux populations.

Je passerai à la présentation du Professeur Silini et à la discussion qui s'en est suivie. Je dirai que cette présentation n'a pas soulevé les mêmes débats, car il a présenté de manière synthétique l'ensemble des données récentes sur les effets biologiques des rayonnements, en soulignant que si les effets déterministes sont maintenant bien fondés, et si les effets stochastiques nécessitent des méthodes d'extrapolation en tenant compte des données épidémiologiques, par contre, tout reste à faire pour les effets héréditaires chez l'homme. Cependant, au cours de la discussion, j'ai été surpris personnellement, je m'implique directement, d'entendre le Professeur Silini nous dire que l'extrapolation à zéro était basée sur des considérations scientifiques, sans parler de l'hormésis, qui, si elle existe, ne pourra jamais être prouvée, ceci revient à dire qu'une seule atteinte à une seule cellule peut suffire à provoquer un cancer. Dans ce cas, on ne pourra pas expliquer au public, et j'avais posé la question hier, pourquoi l'on prend des limites différentes pour le radon qui est naturel et pour les conséquences de l'énergie nucléaire. La discussion sur le mot prudence qui s'en est suivie a montré que nous avons des avis différents sur ce problème et qu'il conviendra certainement d'y revenir. Je soulignerai également que le Professeur Chanteur a fort bien précisé ceci en disant que la relation linéaire permet de prévenir les effets stochastiques, mais pas de les prévoir.

La troisième présentation était celle du Dr. Kouts qui nous a présenté avec beaucoup de clarté comment les gens de la sûreté ont exprimé les objectifs de sûreté, et en particulier, très récemment l'INSAG et nous a proposé de réfléchir d'abord sur les moyens d'atteindre ces objectifs et d'autre part pour vérifier s'ils sont atteints. Dans les deux cas, le seul moyen pour avoir des opinions quantitatives c'est l'utilisation des études probabilistes de sûreté (PSA), mais il faut être bien conscient de leur limitation. C'est là dessus que je voudrais surtout, peut-être, intervenir. Le Dr. Kouts a cité un certain nombre de difficultés inhérentes à l'utilisation des PSA, soit par absence de statistiques sur les accidents, mais ceci est presque une évidence, soit par méconnaissance des phénomènes physiques ou chimiques, notamment dans le développement d'un accident sévère ou encore sur les données pour les événements rares, comme les séismes, M. Uchida nous en a parlé également dans sa présentation. La plus importante des incertitudes, comme l'a bien montré la discussion, c'est la connaissance du facteur humain et des erreurs humaines, la culture de sûreté des équipes, l'organisation du management, varient d'une centrale à l'autre.

Faut-il automatiser pour laisser moins de place aux erreurs humaines, mais alors on devient plus dépendant de la conception des logiciels eux-mêmes et des calculateurs. On ne peut que constater que les deux accidents spectaculaires dans les centrales nucléaires sont dus, pour une grande part, mais pas uniquement, contrairement à ce qu'on dit souvent, à des erreurs des opérateurs, ce qui impose certainement d'accroître fortement la fiabilité humaine, je crois que c'est M. Beninson qui avait insisté là-dessus. M. Högberg a souligné l'importance de ce qu'il a appelé les erreurs humaines subtiles, c'est-à-dire ce qui peut être dû par exemple à la maintenance ou aux essais. Il faut sans doute s'orienter vers des réacteurs plus simples et moins sensibles aux erreurs humaines. Il me semble donc que ce problème du rôle de l'erreur humaine, et plus généralement du facteur humain, est une des limitations dans les études probabilistes. Et d'un autre côté, aujourd'hui même, M. Tanguy a souligné que l'on peut approcher quantitativement ce problème à partir d'essais systématiques sur simulateur, avec des équipes d'opérateurs telles qu'elles fonctionnent dans les centrales, ceci est peut-être plus particulièrement aisé en France avec notre programme très standardisé.

Un dernier point qui a été souligné, la difficulté de la validation des PSA sur l'expérience d'exploitation, l'analyse des incidents, la détection des précurseurs, il faut le faire, et l'AEN a fortement travaillé dans ce sens depuis 10 ans, et ce n'est probablement pas suffisant pour démontrer l'atteinte d'objectifs comme 10^{-6} par an, mais il faut bien retenir que l'utilisation systématique de l'expérience d'exploitation est sans doute aujourd'hui une des voies principales pour améliorer en permanence la sûreté des réacteurs en service, de manière déterministe, même si ceci ne se traduit pas en terme probabiliste direct. Je crois que tout ceci donne quelques éléments que j'ai retenu au cours de cette première session. Merci M. le Président.

K.B. STADIE, NEA

Thank you, Mr. Cogné, for your summary. Perhaps it was a bit more than a summary. It was, to some extent, your personal evaluation of the first session which is, of course, valuable. I think there are people in the room who would probably disagree with some of the observations but we will come back to that in a moment. Let me ask you, Mr. Cogné, before we turn to the next speaker, having chaired the session where we identified the problems, and having then followed through the debate, what do you suggest we do next? Do you believe issues will solve themselves, do we need to discuss matters further, and if so, in what forum and in what way?

F. COGNE, France

Personnellement je pense qu'il faut trouver une méthode pour continuer à dialoguer entre les deux "mafias". Ne serait-ce que pour obtenir une explicitation du dictionnaire, je crois que je reprends les termes du Dr. Gonzalez, je crois qu'il faut trouver des lieux pour le faire. Faut-il en créer de nouveaux, je ne suis pas persuadé, nous sommes tous très pris dans des réunions multiples et variées. Vous avez suggéré tout à l'heure qu'il y ait un peu plus fréquemment un tel séminaire, je pense que c'est utile. Je pense également que l'AEN peut être un bon endroit pour discuter de manière complète et sans que ceci transparaisse à l'extérieur directement, de tous ces différents problèmes. Je crois que nous avons abordé là, j'en ai listé quelques-uns, des points qui certes sont dans les programmes de nos comités, aussi bien au CSNI qu'au CRPPH, mais je pense que ça n'est pas tout à fait suffisant, il y a un besoin de dialogue permanent entre les deux systèmes. Je crois que c'est une des conclusions que je tirerai de cette approche en parallèle des deux systèmes que Mr. Clarke présentait hier. Ce parallèle, il faut qu'il y ait des ponts qui s'établissent, il faut qu'on arrive à expliciter certains termes, certaines approches, et je pense qu'on y reviendra probablement dans la discussion sur un certain nombre de points.

K.B. STADIE, NEA

Thank you, Mr. Cogné. I will refrain from asking others to comment on your summary because we are now turning to Dr. Clarke, who will talk about basic objectives and policies.

R.H. CLARKE, United Kingdom

Thank you, Mr. Chairman. I had the privilege of chairing the second session where you will recall we first took a paper on the ICRP radiation protection philosophy and then three papers on U.S. Safety Goals, European Safety Approaches and the Principal Nuclear Safety Policies of Japan. Let me take the safety issues first. I felt, from the chair, that I detected a convergence of philosophies between the different presentations. It seemed to me that in several areas people from the different countries, different areas of the world were coming to some agreements. Core melt frequencies in the range 10^{-4} to 10^{-5} per year were being mentioned and each of the three papers seemed to propose that accident sequence frequencies of less than 10^{-7} per year were widely regarded as not having to be taken into account. I have deliberately looked in my dictionary and used the word "frequency" because that is what the authors used. There are different measures of the off-site harm as a result of accidents at this agreed frequency level. Sometimes, as in the case of the U.S., we were talking about probabilities of death, early or late deaths associated with that frequency. We had unacceptable radiological consequences from France, for the same frequency, a large uncontrolled release from the U.K., and for Japan, 10^{-7} seemed to be correlated with the need for evacuation or a dose of 0.25 Sv.

So I thought I saw some similar safety goals but technical differences in the way the regulatory authorities treated the consequences. So to anticipate a question you might ask me, chairman, I think here is the first area where I see the need for further progress. And it was certainly suggested in discussion that the industry might be the correct place where the initiative should be taken in working towards some sort of international "harmonisation" to the next stage of safety standards. Regulators, it was said, reflect public concern and they might be pushing for the highest standards to be obtained in this harmonisation of the safety issues. The agreements to which I have referred have been exclusively in terms of individual risk. Professor Uchida specifically addressed the economic consequences of land interdiction, and, in discussion, Commissioner Remick indicated that these other consequences were under consideration in U.S. safety goals. I think this problem is the second one I have identified for future work and so my second point is: Can there be some sort of aggregated measure of harm from an accident? We have said several times that the end points of accidents are so different. You get early deaths, get late deaths at long distances, social disruption from evacuation and relocation, economic losses of produce from land, both agricultural and industrial, and the loss of the land itself.

Can we find some way of expressing all of this harm so that we may use it to search for better safety goals? It should be investigated, but Dr. Beninson said he believed it would never be solved and that may be the case. The alternative approach suggested by Professor Birkhofer is that the next stage in design should aim to make off-site emergency planning unnecessary. So here is my third point. Is that the right objective for nuclear safety goals is that the next generation of reactors should be designed such that off-site consequences basically do not occur. It may still be prudent to have an emergency plan, but the design objective is to ensure that it never needs to be implemented. I think I incline towards this view myself, because it does seem that the social and economic implications of significant off-site releases are basically unacceptable.

And so, finally, to radiation protection. Here the main area of discussion was on intervention and practices, and the reasons for the differences in dose limits for practices and action levels for intervention. Clearly, I have learned that we will have great difficulties in explaining these ideas, but I do think that we now have a single logical framework, distinguishing between situations which add or subtract doses from populations. Previously, we had a system of dose limitation for normal operation, and our advice, from the radiation protection side, had become fragmented. We had given advice on accidents, advice on radon in homes, advice on solid waste disposal - all disappearing off at tangents from the original system of dose limitation. What Dr. Beninson has described to us is one logical framework. It is more coherent, it is logical but you may have difficulty in explaining it to others. I think that this is my final challenge for us all, to explain this new logic.

K.B. STADIE, NEA

Thank you, Dr. Clarke. As usual, clear and concise in your summary. Can I immediately ask you to say a word or two about where do we go from here.

R.H. CLARKE, United Kingdom

Well, I have to agree with Mr. Cogné that meetings of this sort are useful. I think that when we met five years ago the two communities talked at each other and this time the two communities have talked with each other and that is a very big step forward, I genuinely believe. The debate yesterday afternoon between the different communities was extremely valuable and I think profitable, leading me to be able to come to some of the conclusions that I outlined before. I, therefore, think that I would encourage the NEA to have another meeting, not perhaps as far as five years away but two or three years, and I think it would be worth us talking again and reviewing how far we have got with some of these issues that I talked about - the measure of aggregated harm, the acceptability of off-site consequences and where we have got with ICRP.

K.B. STADIE, NEA

Thank you, Dr. Clarke. Now we have covered the fundamental question, the interface, as well as the basic objectives and policies, summarised from representatives of the two Communities, I now open this up to the other Panel members before we come to the achievements in the field.

G. SILINI, Italy

Thank you, Mr. Chairman. I think it would be very easy for me to complain about the fact that the summary of my presentation amounted more to a personal critique by M. Cogné of my arguments than to a cool report of what I said. However, Mr. Chairman, I am really not interested in pursuing this discussion of non-threshold linearity further, I have given my arguments. Let me only point out that I heard no objections raised to the arguments that I gave in answer to Professor Chanteur for assuming non-threshold linearity for the purpose of radiation protection.

K.B. STADIE, NEA

Any other comments to the summaries? Mr. Cogné.

F. COGNE, France

Bien sûr pas sur ce sujet, rassurez-vous. Sur ce que vient de dire Mr. Clarke, il y a un point complémentaire, qui m'a frappé lors de la séance d'hier après-midi, et dont on n'a pas encore parlé, c'est le problème de la perception du public de tout ce que nous disons. J'ai écouté avec beaucoup d'intérêt ce qu'a dit Commissioner Remick, sur le fait qu'aux Etats-Unis la NRC n'a pas réussi à faire comprendre au public ce qu'étaient les objectifs de sûreté et, il a ajouté, ni d'ailleurs aux politiques, donc ceci pose un vrai problème. M. Tanguy avait posé la question d'ailleurs, quelle perception le public a-t-il d'une chance sur un million d'avoir un accident grave. Et je crois que ceci est un problème qui nous interroge, je ne sais pas comment y répondre, il me semble qu'il y a pour la ICRP, mais aussi pour la sûreté, pour tout le monde, une interrogation, car tous les chiffres, tout ce que nous avons utilisé depuis deux jours, il faut bien se rendre compte que nous ne savons pas, ni les uns ni les autres, expliquer au public ni ce que signifie les limites, nous en avons discuté, ni ce que signifie les objectifs de sûreté, c'est ce que disait Commissioner Remick. Je crois que nous avons là un challenge très difficile pour les années qui viennent.

K.B. STADIE, NEA

Can we then turn to the achievements and start with Dr. Beninson.

D. BENINSON, ICRP

I will make a very short summary, but I think that we have achieved in the meeting quite a lot of clarifications that would be extremely beneficial to our work. The main subject in the session I chaired was the question of, on one side what is called potential exposures, and, on the other side, what are accidents and assessment of accidents. I think it became clear that the criterion presented in one of the papers on the so-called potential exposures, i.e. to control the product of probability of occurring times the probability of radiobiological effects given the dose, refers only to controlling individual risk. That means the possibility of attributable death and, therefore, it can be used as a criterion for accidents only in cases where individual risk in that sense is the only consequence of the accident. That is why it has been applied to the cases of irradiation sources. The only consequence there is risk to individuals and, therefore, it applies well and it is not suprising that it does not apply well to the case of reactor accidents where you have a big pot-pourri of consequences and of different magnitudes. Therefore, I think that this was a very important clarification that we had in the meeting.

The second conclusion is that optimisation, which is the same as ALARA, for protection of the population against potential exposures is only applicable reasonably when we are talking of events of not very low probability, say for example 10^{-2} per year and higher, which was the example given by Mr. Tanguy.

The third point that came out of the discussions is that mathematical expectation of consequences, the product of probability times magnitude of the consequence is meaningless when the probability is very low and this, I think, is a very important point which I personally think qualifies, or should qualify, some of the statements that come in probabilistic safety assessments studies. This product seems to be very appealing in some of these assessment and I think it is quite clear that this product, where the probability is very low, is meaningless for decision-making.

If we leave the main subject now, I think there is another subject where there is quite some confusion, and that concerns the applicability of what is called "limit" and what is called "intervention level". This basic confusion, I think, relates to the question why one would limit something at values at which one would not do anything if the radiation is there. I think there is no incorrect basis on that. However, there is a lot of confusion and I think this issue alone, which is quite important for emergency planning, would merit some specific action of NEA - for example a special meeting for discussing the issue and to looking at values that would be reasonable for intervention levels for different types of countermeasures.

It was quite clear from different parts of the presentations that the regulatory practice, and I am not talking of technical issues but of regulatory practice, would really benefit from what is called "exemption levels". That, I think, is a necessity for any regulatory system and it was clear that there is already quite a lot done, by NEA and by the International Atomic Energy Agency, on exemption systems based both on individual dose and collective dose. This does not mean to cut off doses, i.e. to forget about doses below something. It only means to exempt from regulations some practices and some sources.

Finally, we had a different type of discussion regarding the feasibility of using the proposed new dose limits, especially for occupational exposure. The conclusion actually is based not only on what we saw in the session I chaired, but it went over to the next session, and altogether I think one could say that it is feasible to work under the new limits. It would probably not imply an increase of the collective dose, it is likely that it would produce a decrease also of the collective dose, but we identified two problems, one concerning the special types of workers who have been involved in the business for a long time and the problem of underground uranium mining. These two problems are quite real and will require great attention by the industry.

K.B. STADIE, NEA

Thank you Dr. Beninson. You already addressed the question of where we may be of assistance in the dialogue and I will therefore not need to ask you, where we go from here, but turn to Dr. Kouts to complete the summary on achievements and then I will open up the discussion to the other members of the Panel.

H. KOUTS, United States

Dr. Beninson has already introduced one of the papers in the session that I chaired by discussing the problems associated with uranium miners. This was in the first of the two papers that dealt with the effects of the new dose-limits on the very front-end of the fuel cycle. The conclusion there was quite generally, except for the question of the uranium miners, that the new limits really do not raise any undue barriers. The problem in connection with uranium miners does surprise me, because, as we have heard from the Canadian case, and from other cases as well, there seems to exist methods to remove these problems. The practice of using proper ventilation for example, would take us away from the bad experience of many years ago. Altogether the conclusion is pretty well that the new limits that are proposed by ICRP will not cause anybody any great pain.

The second paper in my session had to do with another aspect of trade-off, namely trade-off between safety and radiation protection and there were two aspects discussed in connection with this, whether or not such trade-off actually does take place and the difficulties that would be associated with this trade-off. Mr. Tanguy pointed out, and quite correctly, that in fact there are trade-offs in safety and radiation protection in this field and the principal instance that he cited was that connected with inspection of steam generator tubing, in which you expose individuals so that you can reduce the probability of a large accident in case steam generator tubing were to fail. There are a number of such instances of people being given additional radiation doses during inspection for the purpose of improving overall safety against large accidents in these plants. The interesting matter that was brought up by Mr. Högberg had to do with the difficulties that probabilistic safety analysis brings to the optimisation principle, because he was considering the trade-off between safety and radiation protection as really a trade-off expressed in terms of the optimisation principle as it has been discussed by the ICRP. Here we have a trade-off, an optimisation process, between relatively probable or committed activities in a nuclear power plant and very unlikely events expressed through very low probabilities. The optimisation process one would try to apply in a situation like this is enormously hindered by the difficulties associated with the probabilistic safety analysis itself. That takes us back to old discussions by Reg Farmer many years ago in which he pointed out that there is not the same situation existing in low probability, high consequence events and in high probability, low consequence events. Even though the mathematical expectation values may be the same in two such instances, they reflect by no means the same situation and you therefore have great difficulty in even conceiving of a proper optimisation under such circumstances. That takes us to Mr. Tanguy's paper in which he proposed a cut-off for a probability of about 10^{-2} per year for just this process and this almost closes the circle in the argument.

If I may comment a little further, Mr. Chairman, I would just like to comment on something Dr. Gonzalez said about vocabulary and each of us understanding what the other says. Of course we note that disagreements among people are generally the result of poor communication and what we have been dealing with here is certainly an example. I think that we will not get very far in harmonising our points of view until we begin to harmonise the way we talk about things. Let me propose two terms that require harmonisation. One of them is "risk". We have completely different definitions of the term "risk". We cannot endure this. We are going to have to arrive at one meaning for the term "risk". The second one is "optimisation", which as a term is fundamental in the ICRP methodology. As Mr. Cogné has pointed out, it is not the optimisation we all know, and that we have all had in the mathematics. Optimisation is a strictly mathematical process in which things are made as small as they can possibly be. That is not what you are doing here.

"Optimisation" to me, is a process in calculations in which you minimise the value of a quantity or a function and all of us who have lived in the scientific field all our lives understand that to be meaning of "optimisation". When the term "optimisation" is brought to my attention that is the first thing that leaps to my mind, because I dont know any other meaning of optimisation. I've used it for so many years and I think that is true of many of us. Here are just two problem areas associated with words. Let's see if we can get rid of them.

K. B. STADIE, NEA

Thank you, Dr. Kouts. I now open the summaries for discussion and first give the floor to Dr. Beninson, who wants to make a comment.

D. BENINSON, ICRP

I am not going to solve the problem of "optimisation". I think indeed it requires a discussion on a much more fundamental level than what is the first apparent meaning. I think "optimisation", as used by ICRP, means exactly what you have said in the sense that you either maximise or minimise a given variable which is called the objective variable under some constraints and under some relationships which are usually called the correlating equations and the limiting equations.

K.B. STADIE, NEA

Dr. Kouts has suggested that we get a joint understanding on what "optimisation" means. It is maybe too late now to start the process here, but there are other members of the Panel who have not said anything, so I invite them to enter the discussion on the questions which we have debated so far.

E. GONZALEZ GOMEZ, Spain

I just wanted to make a very small reflection and stress something that has already been said. Dr. Clarke has said that in this meeting we have talked with each other and not only at each other. I agree completely with that and I think that if we look at the areas of discussion, the things that need further analysis and that need some convergence between, both activities, I really do not see that much of a difference between, what has been called during the meeting, both communities. What I mean is that when there are problems between what is called the safety community and the radiation protection community, it is often related to the fact that the reactor safety community finds problems with public opinion in many of the aspects that are raised by radiation protection activities. This is at least the case in regulatory organisations, that have to deal with both sides of the coin. I think the problem is much more how to convey the message to the public, rather than to discuss among ourselves, so I agree with Dr. Bengtsson when he said that the problem of communication is a very important one and we will have to try to work more with each other than against each other.

K.B. STADIE, NEA

A point well taken, Dr. Gonzalez Gomez. Professor Uchida.

H. UCHIDA, Japan

Thank you, Mr. Chairman. First of all I would like to summarise again the basic safety principles in Japan which are in my paper but were not described at yesterday's presentation. The current Japanese radiation protection law was revised in 1989, being based on ICRP Publication 26. The maximum permissible dose for a person of the public is 1 mSv per year. However, it was decided, according to the ALARA principle, that the radiation exposure to the surrounding regions in the vicinity of a nuclear power station during normal operation shall not exceed 50 µSv per year per site. In practice, the actual exposure level for the public is estimated to be less than one tenth of the figure mentioned above, i.e. about 5 to 1 µSv per year per site.

An important prerequisite for countermeasures against accident consequences is that the population in surrounding regions do not receive non-stochastic effects of radiation, even in the case of a highly unlikely accident. For this purpose the safety of a nuclear power plant is ensured by proper design, manufacture, construction and operation, based on high level technology and implementation of preventive maintenance. Safety of a nuclear power plant should be ensured entirely within each site. No off-site emergency procedure should be necessary. Emergency countermeasures outside the site do refer to actions to be taken as precaution by administrative organisations. These operations would play a partial role in accident management in mitigating the off-site radiological consequences and residual risks of a severe accident which leads to a significant release of radioactive material. Countermeasures for emergencies concerning nuclear accidents in Japan are based on the Fundamental Law for Disaster Management which is applied to earthquakes, typhoons, tidal waves and all other disasters. This law prescribes that the administrative organisations should be responsible for establishing emergency procedures and implementing them when necessary.

The dose to workers during maintenance and periodic inspections is presently 2 to 3 mSv per year, as an average. There are a few key groups of workers in nuclear facilities who are exposed to more than 20 mSv per year. The collective dose for workers in nuclear power plants due to periodic inspection which, in accordance with the law, is conducted once a year, is decreasing year after year and the figure is now around 2 to 5 manSv per reactor and year.

I have appreciated very much the interesting discussions which have taken place in this meeting. They have been very useful in understanding the international trend in searching for a framework of the entire safety problem.

K.B. STADIE, NEA

Thank you, Professor Uchida. Let me give the floor to Dr. Bengtsson and then I will open the contributions to the discussions.

G. BENGTSSON, Sweden

I would like to hang on to what Dr. Gongalez Gomez said about the communication. In order to get optimal management of nuclear risks we must be able to communicate with the general public and with the politicians, and one thing which has been lacking in our discussions here is that we must not focus entirely on our own risks but we must also provide a perspective on other risks that are outside of our field. I think the time is very ripe for pursuing such a perspective. I will give you an example from Sweden, where we have had a government investigation two years ago, looking at nuclear emergency planning, but where for the first time in Swedish investigations, they were also looking at other kinds of risks in the same context, with the same people. They were also looking at such things as chlorine transports or chemical factories' emergency management. I think the time has come to discuss this and I would like to suggest that if one is developing in a Seminar our common dictionary, one should let this dictionary deal not only with the nuclear risks but also with other risks and discuss how we present our (nuclear) risks and how we present other risks, i.e. the complete risk scenario which exists in our society.

K.B. STADIE, NEA

Your point is well taken Dr. Bengtsson. I do believe that we need to ultimately communicate with the outside world. In fact our credibility critically depends on public acceptance of nuclear power and its associated risks but, after our two days of discussions here, I think we should get our house in order before we try to address the public. There are obviously differences which we need to resolve. Dr. Kouts has pointed out two fundamental terms on which we apparently have some differences of opinion on what they mean.

I will now turn to Dr. Beninson, who has a question.

D. BENINSON, ICRP

It is a very narrow question to Professor Uchida. If I understood correctly yesterday, it is not possible to apply a probabilistic approach in the case of earthquakes in Japan because of specific geological conditions.

H. UCHIDA, Japan

Our design policy as regards earthquakes is basically the following:

For regulatory design purposes large natural forces are expected to occur in the vicinity of the site of a nuclear power plant. These incude earthquakes, tsunamis (high sea waves), typhoons, and floods. The basic approach to nuclear power plant design is to assume natural forces larger than the largest events experienced at the vicinity of the site in history. This concept is described in the IAEA Safety Series No. 50-C-S as "the radiological risks associated with external events should not exceed the severity of radiological risks associated with accidents of internal origin". In such design considerations, the seismic design against earthquakes must have the greatest importance in Japan.

A more rigorous approach is taken in designs when addressing seismic forces than when addressing other natural forces. The sites of nuclear power plants in Japan are selected at seismically stabilized zones which are distant from highly active faults. Geologically hazardous areas are excluded. The foundation of a nuclear power plant is constructed on firm bedrock which originated in the Tertiary geological era or earlier.

The seismic design is performed so that a reactor accident which would lead to large releases of radioactive materials could not be caused by either the largest historical earthquake or a much more severe earthquake which is extremely unlikely to occur but can be postulated from engineering observations based on seismotectonics, etc. An earthquake ground motion which could shake the reactor facilities by an earthquake postulated for this purpose is called a basic design earthquake ground motion. Two types of basic earthquake ground motions are postulated: an earthquake ground motion S1 caused by the strongest earthquake considered to have a possibility of occurring from an engineering viewpoint (called the maximum design basis earthquake) and an earthquake ground motion S2 caused by the strongest earthquake which can be assumed to take place based on the engineering observations of the seismotectonics around the site, etc. (called the extreme design basis earthquake).

The basic earthquake ground motion is defined based on the design earth-quake. The magnitude and location are postulated, taking into account the seismotectonics, active faults associated with earthquakes, the historical records of earthquakes which occurred in the region surrounding the site, etc. The basic earthquake ground motion is expressed by a response spectrum or a time-history simulated seismic wave with maximum amplitude, frequency charac-teristics, durations and changes with time taken into consideration. (The maximum amplitude of the earthquake ground motion is normally indicated by a velocity.)

Historical records are surveyed for descriptions of earthquakes that occurred within about 200 km of the site. Those earthquakes of Seismic Intensity V (called a strong earthquake) or higher by the Meteorological Agency's intensity scale (Intensity 0-VII) are examined closely. Intensity V is equivalent to an Intensity of 7 3/4 on the Modified Mercalli Intensity Scale. An extreme design earthquake associated with active fault displacement in the vicinity of the site which is postulated to occur from seismotectonic considerations usually ranges to Magnitude 7.5-8.5. For an extreme design earthquake, an earthquake of at least Magnitude 6.5 is postulated to occur about 10 km directly under the reactor facility centre.

From investigation and research of historical earthquake records, seis-mology and the seismotectonic structure of Japan, the following concept is drawn: an earthquake will occur by releasing the energy accumulated within the zone concerned. The zone will have an infinite area, and the recurrent period which is inherent to the zone will be some thousands of years or less. From this concept it is concluded that there is an upper boundary for the maximum magnitude of the potential maximum earthquake estimated to occur in the vicini-ty of sites in Japan. An extreme design earthquake corresponds to the poten-tial maximum earthquake mentioned above or is larger. It seems unreasonable to assume, based on so-called probabilistic analyses, that a larger earthquake would occur with a recurrent period greater than some thousands of years.

K.B. STADIE, NEA

Thank you, Prof. Uchida.

H. KOUTS, United States

I want to go back to something that Dr. Bengtsson said concerning wideness of concept of risk and plead that, in this conciliation that we are moving towards here, we bear in mind the point that he has made. Let us do this with a sense of proportion and I am reminded of a book by Blasco Ibanez, the Spaniard, called "The Four Horsemen of the Apocalypse". You may have heard of it. The four horsemen appear in the Bible and are War, Famine, Disease and Death and I don't think we want to give the public the impression that we are adding a fifth horseman of the apocalypse in the form of radiation. We should keep a sense of proportion here. We are faced with a job which we want to accomplish, and it is an important job, but it ranks where it ranks and we don't rank it at the very top. We should join it with all the other things which have to be optimised.

K.B. STADIE, NEA

Thank you, Dr. Kouts. Mr. Bonn, please.

M. BONN, Financial Times

I just wanted to ask you, given that there is such an enormous misconception about radiation and nuclear risk on the part of the public in our countries, what could each of your communities, as you call them, contribute to lessening this misconception on the part of the public?

D. BENINSON, ICRP

I don't think that we should try to cure the misconception. I think the misconception will be cured by itself when the public sees what goes on. The less we try to influence the public, the better, and that is my personal feeling. When we talk about public comprehension, communication and so on, I sometimes have a feeling that what we really mean is propaganda and I don't think we should do any of this propaganda. I think that the public is clever enough to see and when they have discussed the issues they eventually come up with a clear cut opinion. What distorts this opinion is not that we communicate well or not, but that different political figures use the issue for very different purposes so perhaps you should address the question more to the politicians.

F. COGNE, France

Ce que je voulais dire simplement, c'est un petit peu dans le même sens. Il est important que le public actuellement ne fait pas la distinction entre la protection et la sûreté. Quand on parle de normes, il met aussi bien dedans les normes de radioprotection que les normes de sûreté. Il ne sait pas de quoi il est question. Je pense que cette confusion nous l'entretenons sans doute, et on a des exemples en ce moment en France, tout à fait évidts, et je crois qu'il y a un effort d'explication, de compréhension qui me paraît tout à fait important. Je ne suis pas tout à fait d'accord avec M. Beninson, il ne suffit pas de s'en remettre aux politiques, on a tous en tant que scientifiques, ingénieurs, une responsabilité à faire comprendre de quoi nous parlons, ce que nous voulons dire et d'être aussi clairs que possible. Je crois qu'on ne peut qu'être un peu, j'allais dire pessimistes, ou désappointés qu'on n'ait pas réussi à se faire comprendre aussi bien, mais je crois qu'il y a un effort permanent qui est à faire dans ce domaine, et je ne crois pas qu'il faille se réfugier dans les politiques. Nous nous devons d'expliquer ce que c'est que des normes de sûreté, des normes de protection. Admettre que ce ne sont pas les mêmes, nous en avons assez discuté, mais à ce moment-là il faut bien l'expliciter, ce qu'on a dit tout à l'heure, ce qu'a souligné M. Kouts sur le problème d'harmonisation des termes, sur les risques, l'optimisation, savoir de quoi on parle, qu'est-ce que c'est que les risques en sûreté, les risques en protection, je crois qu'il y a un effort à faire.

K.B. STADIE, NEA

Your question, Mr. Bonn, has obviously incited the Panel. Next, Professor Silini and then Dr. Bengtsson.

G. SILINI, Italy

Yes, Mr. Chairman, it is a very interesting question and if I could answer to it from my own point of view, I would say that my recipe would be to be as open as possible, always. And when there are problems, point them out, recognize them and tell people how we go about solving them and the difficulties, the agonising difficulties sometimes, of having to take decisions under certain circumstances; and this should be done at all levels, from the public to the politicians, to everybody - and that I think is the only way, in my own view, to solve the problem. It is not by telling half-truths that you go about solving the problem, because the policy of half-truths is the policy of half-lies.

G. BENGTSSON, Sweden

If I may continue, I would advocate the recipe I already advocated - to devote significant efforts to establishing a similar basis for the management of risks in different environmental areas and I gave you before the example of emergency management where we have had the practical example in Sweden. We have other practical examples. For instance, our Environmental Protection Board has established a linearity principle for chemical carcinogenics just as for radiation and we are having, next year, a large international symposium on the management of all kinds of waste - hazardous waste - whether it be radioactive or not. I think the time is ripe for a large number of such efforts and when the general public comes to see different hazardous agents side by side they will eventually make their own decisions on the hazards of nuclear power and my guess is that in most instances they will realize that so far many other judgments have been exaggerated.

K.B. STADIE, NEA

Thank you, Dr. Bengtsson. Dr. Clarke.

R.H. CLARKE, United Kingdom

I am going to agree with what other people have said. The solution cannot be simple, otherwise we would already have found it, wouldn't we, and it is not a question that is unique to radiation, it is a question of technology in modern life. I think it is a general question of the way society handles things which have great benefits but small and quantifiable risks and society has not found a way to manage that, yet. We have to be honest, avoid propaganda and give information, of course. I don't think that we can do any more than that, but it is interesting to reflect in our own field in radiation protection, the great difficulty we have in persuading people with high radon levels in their own homes to take some action about it. People do not like to think that there is something in their lifestyle or something in their genes which adversely affects them. They would much rather blame "them" – that is the operator of a power station or a rubbish dump or another airport at the end of the garden. People want to blame "them", so I say it is something to do with human nature, it is something to do with technology in society and I honestly think it will take a generation to get through to this, because we have to educate people, people have to do the balances that Dr. Bengtsson wanted and have to do it in probabilistic terms and, although people back horses and do the football pools, they do not really understand probabilities. But, at some point in the future they have to be able to do that in order to make the judgments that Dr. Bengtsson said.

K.B. STADIE, NEA

Time is running on. A last remark from Dr. Gongalez.

A. GONZALEZ, IAEA

It has been said already that openness in the transmission of information is an important issue. Another one, perhaps, is not to give the impression to the public that we are transmitting our technical responsibilities to them, because this has nothing to do with openness, this is pure demagogy. The technical problems in this field are our problems. We have the technical responsibility to solve these problems and this happens in any field of technical life. I am also a member of the public and I am not interested in participating in the definition of the safety features of the plane that I use to fly in for the very simple reason that I do not understand aeronautic engineering. There are people who have that responsibility. I pay salaries indirectly to them to solve that problem. I want to be informed, but I do not want that they share their responsibilities with me and we should be very careful with that, because today in some of the discussions there was a flavour, particularly by my Swedish colleague, to pass the ball to the public to take decisions. This is a very big mistake, I believe, and together with a lack of openness in information, it can produce a very negative effect on the public.

K.B. STADIE, NEA

Well, I think we went from the general to the specific and we must return to the very general so maybe this was a good way to close the circle. I would now think it is time to close the meeting. I will try to summarise in very few words because it is impossible to even begin to give a full account of what has happened over the last two days. I think we all agree that it was a very dynamic meeting and I agree with Dr. Clarke that we this time talked with each other. I think the problem we are facing is that we are living in a world which is probabilistic in nature and a world which has to make regulatory decisions which are, by definition, binary in nature, in a world where people like to have electricity but are not willing to accept the risks that come with it. That is one side of the problem. The other is that, as I said at the beginning, I believe that nuclear industry is uncommonly honest in trying to address the risks it poses to mankind but we should not go too far in the probabilistic ranges to cover areas which we do not understand well enough. I only refer here to Reg Farmer who reminds us more and more frequently that the low probability events have no physical meaning. I will leave it at that and will try to indicate briefly what I conclude we go from here.

Several people have talked to me during the coffee breaks and suggested that we have to have meetings like this every six months. I think it is impossible for our agency to sponsor such an exercise very frequently and therefore I like much better the idea by Dr. Clarke that we meet in, let's say three years time. I also believe that we have reached a point in our dialogue where it is possible to sub-divide the issues into smaller parcels and take them up one by one. I would like to remind you that we have three committees at NEA which cover radiation protection, the CRPPH, nuclear safety technology and nuclear safety issues, the CSNI, and regulatory aspects of reactor safety, the CNRA.

Maybe our Agency can help through these three committees, in finding a framework in which we can tackle some of the remaining questions, beginning perhaps with the vocabulary, as was suggested, and Dr. Kouts has already given us one or two suggestions where to begin.

If we follow such a course, I do believe that we will now make rapid progress and once we have our house in order, I think we have the task to explain to our compatriots what we understand but also, and I agree with Dr. Bengtsson, what we do not understand. We have to clearly define our limits and people have to understand that zero risk is not possible, and approaching zero risk is maybe economically unwise to aim at. Thank you.

AUSTRALIA - AUSTRALIE

MURRAY, A., Nuclear Safety Bureau, Australian Nuclear Science & Technology
Organisation (ANSTO), P.O. Box 346, Menai, New South Wales 2234

BELGIUM - BELGIQUE

GOVAERTS, P., Département d'Etudes de Sûreté Nucléaire, AV Nucléaire,
157 avenue du Roi, 1060 Bruxelles

STALLAERT, P.E., Inspecteur général, Service de la Sécurité Technique des
Installations Nucléaires (SSTIN), Ministère de l'Emploi et du Travail,
51 rue Belliard, 1040 Bruxelles

CANADA

BOND, J.A., Manager, Radiation & Industrial Safety Branch, Occupational Health
and Safety Division, AECL, Research Company, Chalk River Laboratories,
Chalk River, Ontario K0J 1J0

DOMARATZKI, Z., Director General, Atomic Energy Control Board, P.O. Box 1046,
Ottawa K1P 5S9

DUNCAN, R.M., Manager, Radiation Protection Division, Atomic Energy Control
Board, P.O. Box 1046, Ottawa K1P 5S9

FINLAND - FINLANDE

SALO, A., Director of Surveillance Department, Finnish Centre for Radiation and
Nuclear Safety, P.O. Box 268, SF-00101 Helsinki

TOIVOLA, A., Manager of Nuclear Safety, Teollisuuden Voima Oy,
SF-27160 Olkiluoto

VUORINEN, A.P.U., Director General, Finnish Centre for Radiation and Nuclear
Safety, Elimäenkatu 25 A, 00510 Helsinki

FRANCE

CHANTEUR, J., Directeur Adjoint du Service Central de Protection contre les Rayonnements Ionisants (SCPRI), Ministère des Affaires Sociales et de la Solidarité Nationale, B.P. n° 35, 78110 Le Vésinet

COGNE, F., Inspecteur général pour la sûreté nucléaire, Commissariat à l'Energie Atomique, Centre d'Etudes Nucléaires de Fontenay-aux-Roses, B.P. n° 6, F-92265 Fontenay-aux-Roses Cedex

COULON, R.B., Adjoint au Chef du Département de Protection Sanitaire, IPSN/DPS, Commissariat à l'Energie Atomique, B.P. n° 6, 92265 Fontenay-aux-Roses Cedex

FITOUSSI, L., Conseiller Scientifique, Service Central de Protection contre les Rayonnements Ionisants, Ministère des Affaires Sociales et de la Solidarité Nationale, Département des Sciences Nucléaires, Conservatoire National des Arts et Métiers (CNAM), Bâtiment 476, CEN Saclay, 91191 Gif-sur-Yvette Cedex

HENRY, P.H., COGEMA, B.P. 4, 78 141 Velizy Villacoublay Cedex

HULST, J., Service Central de Sûreté des Installations Nucléaires (SCSIN), Ministère de l'Industrie, 99 rue de Grenelle, 75007 Paris

LAVERIE, M., Chef du Service Central de Sûreté des Installations Nucléaires, Ministère de l'Industrie et de l'Aménagement du Territoire, 99 rue de Grenelle, F-75700 Paris Cedex

L'HOMME, A., IPSN/DAS, Commissariat à l'Energie Atomique, Centre d'Etudes Nucléaires de Fontenay-aux-Roses, Boîte Postale 6, F-92265 Fontenay-aux-Roses Cedex

TANGUY, P.Y., Inspecteur Général, Electricité de France, 32 rue de Monceau, 75384 Paris Cedex 08

GERMANY - ALLEMAGNE

BIRKHOFER, A., Managing Director, Lehrstuhl für Reaktordynamik und Reaktorsicherheit der Technischen, Universität München, Gesellschaft für Reaktorsicherheit mbH, Forschungsgelände, D-8046 Garching

GAST, K., Ministerialdirigent, Director, Safety of Nuclear Installations, Bundesministerium für Umwelt, Naturschutz und Reaktorsicherheit (BMU), Husarenstrasse 30, Postfach 120629, D-5300 Bonn 1

HARDT, H.-J., Federal Ministry of Environment, Nature Conservation and Reactor Safety, Husarenstrasse 30, D-5300 Bonn

JAHNS, A., (GRS), c/o IPSN-DRSN, Centre d'Etudes Nucléaires de Fontenay-aux-Roses, B.P. n° 6, 92265 Fontenay-aux-Roses Cedex

IRELAND - IRLANDE

TURVEY, F.J., Assistant Chief Executive, Nuclear Energy Board, 3 Clonskeagh
   Square, Clonskeagh Road, Dublin 14

ITALY - ITALIE

BENASSAI, S., Assistant to Director of Division of Standards, ENEA/DISP, Via
   Vitaliano Brancati 48, I-00144 Rome

FRULLANI, S., Istituto Superiore di Sanita, Viale Regina Elena 299, 00161 Rome

NASCHI, G., Director, Directorate of Nuclear Safety and Health Protection, Via
   Vitaliano Brancati 48, 00144 Rome

SILINI, G., Via S. Maurizio 22, 24065 Lovere (Bergamo)

SUSANNA, A.F., Direttore Settore Ambiente e Radioprotezione, ENEA/DISP, Via
   Vitaliano Brancati 48, 00144 Rome

JAPAN - JAPON

KONDO, S., Senior Engineer, O-arai Engineering Center, Power Reactor and
   Nuclear Development Corporation, 4002 Narita, O-arai, Higashi-Ibaraki-Gun,
   Ibaraki 311-13

MAEZAWA, Y., First Secretary, Delegation of Japan to the OECD, 7 avenue Hoche,
   75008 Paris

OSHINO, M., Director, Department of Health Physics, JAERI, Tokai Research
   Establishment, 2-4 Shirane, Shirakata, Tokai-mura, Naka-gun, Ibaraki-ken

TANI, H., Director, Nuclear Safety Policy Division, Nuclear Safety Bureau,
   Science and Technology Agency, 2-2-1 Kasumigaseki, Chiyoda-ku, Tokyo 100

TOBIOKA, T., Deputy Director, Department of Reactor Safety Research, Tokai
   Research Establishment, JAERI, Tokai-mura, Naka-gun, Ibaraki-ken 319-11

UCHIDA, H., Chairman, Nuclear Safety Commission, Science and Technology Agency,
   2-2-1 Kasumigaseki, Chiyoda-ku, Tokyo

YOSHIDA, T., Director, Radiation Protection Division, Science and Technology
   Agency, 2-2-1 Kasumigaseki, Chiyoda-ku, Tokyo

NETHERLANDS – PAYS-BAS

KEVERLING BUISMAN, A.S., Energy Research Foundation (ECN), P.O. Box 1, NL-1755 ZG Petten

DE MUNK, P., Ministry of Social Affairs and Employment, Nuclear Safety Department, P.O. Box 69, 2270 MA Voorburg

NORWAY – NORVEGE

BAARLI, J., Director, National Institute of Radiation Hygiene, Box 55, 1345 Österas

GUSSGARD, K., Director, Norwegian Nuclear Energy, Safety Authority, P.B. 750 Sentrum, N-0106 OSLO 1

PORTUGAL

MARQUES de CARVALHO, A., Gabinete de Proteçcao e Segurança Nuclear, Avenida da Republica 45-6°, P-1000 Lisbon

SPAIN – ESPAGNE

AMOR, I., Consejo de Seguridad Nuclear, Justo Dorado 11, 28040 Madrid

GONZALEZ GOMEZ, E., Vice-President, Consejo de Seguridad Nuclear, c/Sor Angela de la Cruz 3, 28020 Madrid

IRANZO, E., Centro de Investigaciones Energeticas Medioambientales y Tecnologicas (CIEMAT), Avenida Complutense 22, 28040 Madrid

SWEDEN – SUEDE

BENGTSSON, L.G., Director General, Swedish Radiation Protection Institute, Box 60204, S-10401 Stockholm

HÖGBERG, L.G., Director General, Swedish Nuclear Power Inspectorate, Box 27106, S-102 52 Stockholm

PERSSON, Å., Swedish Radiation Protection Institute (SSI), Box 60204, S-104 01 Stockholm

SNIHS, J.O., Swedish Radiation Protection Institute (SSI), Box 60204,
S-104 01 Stockholm

SWITZERLAND - SUISSE

GONEN, Y.G., Swiss Federal Nuclear Safety Inspectorate (HSK),
CH-5303 Würenlingen

MICHAUD, B., Office Fédéral de la Santé Publique, Bollwerk 27, Case
Postale 2644, CH-3001 Berne

NAEGELIN, R., Director, Swiss Federal Nuclear Safety Inspectorate,
CH-5303 Würenlingen

PRETRE, S.B., Head, Division of Radiation Protection, Nuclear Safety
Inspectorate, CH-5303 Würenlingen

UNITED KINGDOM - ROYAUME-UNI

BEAVER, P.F., Health & Safety Executive, Baynards House, 1 Chepstow Place,
London W2 4TF

BERRY, R.J., Director of Health, Safety and Environmental Protection, Britis
Nuclear Fuels plc, Risley, Warrington, Cheshire WA3 6AS

CAMPBELL, H.E., UKAEA (AEA Technology), Safety Directorate, Wigshaw Lane,
Culcheth, Cheshire, WA3 4NE

CAMPBELL, J.F., H.M. Superintending Inspector, H.M. Nuclear Installations
Inspectorate, St. Peter's House, Stanley Precinct, Bootle, Merseyside

CLARKE, R.H., Director, National Radiological Protection Board, Chilton,
Didcot, Oxfordshire OX11 ORQ

REED, J.M., H.M. Superintending Inspector, H.M. Nuclear Installations
Inspectorate, Baynards House, 1 Chepstow Place, Westbourne Grove,
London W2 4TF

RYDER, E.A., H.M. Chief Inspector, H.M. Nuclear Installations Inspectorate,
Health & Safety Inspectorate, Baynards House, 1 Chepstow Place,
London W2 4TF

UNITED STATES - ETATS-UNIS

CUNNINGHAM, R.E., Director, Division of Industrial and Medical Nuclear Safety, Office of Nuclear Material Safety and Safeguards, U.S. Nuclear Regulatory Commission, Washington DC 20555

FAY, C.W., Vice President, Nuclear Power, Wisconsin Electric Power Company, 231 West Michigan Street, Milwaukee, Wisconsin 53201

GUTTMANN, J., Technical Assistant to the Commissioner, Nuclear Regulatory Commission, Washington, D.C. 20555

KOUTS, H., Defense Nuclear Facilities', Safety Board, 600 E Street, Suite 675, Washington DC 20004

MacDOUGALL, R., Technical Assistant to the Commissioner, Nuclear Regulatory Commission, Washington, D.C. 20555

REMICK, F.J., Commissioner, Nuclear Regulatory Commission, Washington DC 20555

SISSON, J., Administrative Assistant to the Commissioner, Nuclear Regulatory Commission, Washington, D.C. 20555

COMMISSION OF THE EUROPEAN COMMUNITIES
COMMISSION DES COMMUNAUTES EUROPEENNES

CIANI, V., Directorate-General Environment, Nuclear Safety and Civil Protection, XI-A-1, Radiation Protection, Commission of the European Communities, 200 rue de la Loi, B-1049 Brussels, Belgium

FINZI, S., Commission of the European Communities, 200 rue de la Loi, B-1049 Brussels, Belgium

KELLY, G.N., Commission of the European Communities, 200 rue de la Loi, B-1049 Brussels, Belgium

ORLOWSKI, S., Commission of the European Communities, 200 rue de la Loi, B-1049 Brussels, Belgium

INTERNATIONAL ATOMIC ENERGY AGENCY
AGENCE INTERNATIONALE DE L'ENERGIE ATOMIQUE

GONZALEZ, A.J., Deputy Director, Division of Nuclear Safety, Head, Radiation Protection Section, International Atomic Energy Agency, P.O. Box 100, A-1400 Vienna, Austria

INTERNATIONAL COMMISSION ON RADIOLOGICAL PROTECTION
COMMISSION INTERNATIONALE DE PROTECTION CONTRE LES RADIATIONS

BENINSON, D., Chairman, International Commission on Radiological Protection, CNEA, Ave. Libertador 8250, 1429 Buenos Aires, Argentina

INTERNATIONAL RADIATION PROTECTION ASSOCIATION
ASSOCIATION INTERNATIONALE POUR LA PROTECTION CONTRE LES RADIATIONS

UZZAN, G., IPSN, Membre du Comité Exécutif de l'IRPA, Commissariat à l'Energie Atomique, B.P. n° 6, 92265 Fontenay-aux-Roses Cedex, France

OECD NUCLEAR ENERGY AGENCY
AGENCE DE L'OCDE POUR L'ENERGIE NUCLEAIRE

GALLIOT, F., Information Officer, OECD Nuclear Energy Agency, 38 boulevard Suchet, F-75016 Paris (France)

ILARI, O., Division of Radiation Protection and Waste Management, OECD Nuclear Energy Agency, 38 boulevard Suchet, F-75016 Paris (France)

McPHERSON, G.D., Division of Nuclear Safety, OECD Nuclear Energy Agency, 38 boulevard Suchet, F-75016 Paris (France)

STADIE, K., Deputy Director, Safety and Regulation, OECD Nuclear Energy Agency, 38 boulevard Suchet, F-75016 Paris (France)

UEMATSU, K., Director General, OECD Nuclear Energy Agency, 38 boulevard Suchet, F-75016 Paris (France)

VIKTORSSON, C., Division of Radiation Protection and Waste Management, OECD Nuclear Energy Agency, 38 boulevard Suchet, F-75016 Paris (France)

WHERE TO OBTAIN OECD PUBLICATIONS – OÙ OBTENIR LES PUBLICATIONS DE L'OCDE

Argentina – Argentine
CARLOS HIRSCH S.R.L.
Galería Güemes, Florida 165, 4° Piso
1333 Buenos Aires   Tel. 30.7122, 331.1787 y 331.2391
Telegram: Hirsch-Baires
Telex: 21112 UAPE-AR. Ref. s/2901
Telefax:(1)331-1787

Australia – Australie
D.A. Book (Aust.) Pty. Ltd.
648 Whitehorse Road, P.O.B 163
Mitcham, Victoria 3132   Tel. (03)873.4411
Telefax: (03)873.5679

Austria – Autriche
OECD Publications and Information Centre
Schedestrasse 7
D-W 5300 Bonn 1 (Germany)   Tel. (49.228)21.60.45
Telefax: (49.228)26.11.04
Gerold & Co.
Graben 31
Wien I   Tel. (0222)533.50.14

Belgium – Belgique
Jean De Lannoy
Avenue du Roi 202
B-1060 Bruxelles   Tel. (02)538.51.69/538.08.41
Telex: 63220   Telefax: (02) 538.08.41

Canada
Renouf Publishing Company Ltd.
1294 Algoma Road
Ottawa, ON K1B 3W8   Tel. (613)741.4333
Telex: 053-4783   Telefax: (613)741.5439
Stores:
61 Sparks Street
Ottawa, ON K1P 5R1   Tel. (613)238.8985
211 Yonge Street
Toronto, ON M5B 1M4   Tel. (416)363.3171
Federal Publications
165 University Avenue
Toronto, ON M5H 3B8   Tel. (416)581.1552
Telefax: (416)581.1743
Les Publications Fédérales
1185 rue de l'Université
Montréal, PQ H3B 3A7   Tel.(514)954-1633
Les Éditions La Liberté Inc.
3020 Chemin Sainte-Foy
Sainte-Foy, PQ G1X 3V6   Tel. (418)658.3763
Telefax: (418)658.3763

Denmark – Danemark
Munksgaard Export and Subscription Service
35, Nørre Søgade, P.O. Box 2148
DK-1016 København K   Tel. (45 33)12.85.70
Telex: 19431 MUNKS DK   Telefax: (45 33)12.93.87

Finland – Finlande
Akateeminen Kirjakauppa
Keskuskatu 1, P.O. Box 128
00100 Helsinki   Tel. (358 0)12141
Telex: 125080   Telefax: (358 0)121.4441

France
OECD/OCDE
Mail Orders/Commandes par correspondance:
2, rue André-Pascal
75775 Paris Cédex 16   Tel. (33-1)45.24.82.00
Bookshop/Librairie:
33, rue Octave-Feuillet
75016 Paris   Tel. (33-1)45.24.81.67
    (33-1)45.24.81.81
Telex: 620 160 OCDE
Telefax: (33-1)45.24.85.00 (33-1)45.24.81.76
Librairie de l'Université
12a, rue Nazareth
13100 Aix-en-Provence   Tel. 42.26.18.08
Telefax : 42.26.63.26

Germany – Allemagne
OECD Publications and Information Centre
Schedestrasse 7
D-W 5300 Bonn 1   Tel. (0228)21.60.45
Telefax: (0228)26.11.04

Greece – Grèce
Librairie Kauffmann
28 rue du Stade
105 64 Athens   Tel. 322.21.60
Telex: 218187 LIKA Gr

Hong Kong
Swindon Book Co. Ltd.
13 - 15 Lock Road
Kowloon, Hong Kong   Tel. 366.80.31
Telex: 50 441 SWIN HX   Telefax: 739.49.75

Iceland – Islande
Mál Mog Menning
Laugavegi 18, Pósthólf 392
121 Reykjavik   Tel. 15199/24240

India – Inde
Oxford Book and Stationery Co.
Scindia House
New Delhi 110001   Tel. 331.5896/5308
Telex: 31 61990 AM IN
Telefax: (11)332.5993
17 Park Street
Calcutta 700016   Tel. 240832

Indonesia – Indonésie
Pdii-Lipi
P.O. Box 269/JKSMG/88
Jakarta 12790   Tel. 583467
Telex: 62 875

Ireland – Irlande
TDC Publishers – Library Suppliers
12 North Frederick Street
Dublin 1   Tel. 744835/749677
Telex: 33530 TDCP EI   Telefax: 748416

Italy – Italie
Libreria Commissionaria Sansoni
Via Benedetto Fortini, 120/10
Casella Post. 552
50125 Firenze   Tel. (055)64.54.15
Telex: 570466   Telefax: (055)64.12.57
Via Bartolini 29
20155 Milano   Tel. 36.50.83
La diffusione delle pubblicazioni OCSE viene assicurata
dalle principali librerie ed anche da:
Editrice e Libreria Herder
Piazza Montecitorio 120
00186 Roma   Tel. 679.46.28
Telex: NATEL I 621427
Libreria Hoepli
Via Hoepli 5
20121 Milano   Tel. 86.54.46
Telex: 31.33.95   Telefax: (02)805.28.86
Libreria Scientifica
Dott. Lucio de Biasio 'Aeiou'
Via Meravigli 16
20123 Milano   Tel. 805.68.98
Telefax: 800175

Japan – Japon
OECD Publications and Information Centre
Landic Akasaka Building
2-3-4 Akasaka, Minato-ku
Tokyo 107   Tel. (81.3)3586.2016
Telex: (81.3)3584.7929

Korea – Corée
Kyobo Book Centre Co. Ltd.
P.O. Box 1658, Kwang Hwa Moon
Seoul   Tel. (REP)730.78.91
Telefax: 735.0030

Malaysia/Singapore – Malaisie/Singapour
Co-operative Bookshop Ltd.
University of Malaya
P.O. Box 1127, Jalan Pantai Baru
59700 Kuala Lumpur
Malaysia   Tel. 756.5000/756.5425
Telefax: 757.3661
Information Publications Pte. Ltd.
Pei-Fu Industrial Building
24 New Industrial Road No. 02-06
Singapore 1953   Tel. 283.1786/283.1798
Telefax: 284.8875

Netherlands – Pays-Bas
SDU Uitgeverij
Christoffel Plantijnstraat 2
Postbus 20014
2500 EA's-Gravenhage   Tel. (070 3)78.99.11
Voor bestellingen: Tel. (070 3)78.98.80
Telex: 32486 stdru   Telefax: (070 3)47.63.51

New Zealand – Nouvelle-Zélande
GP Publications Ltd.
Customer Services
33 The Esplanade - P.O. Box 38-900
Petone, Wellington
Tel. (04)685-555   Telefax: (04)685-333

Norway – Norvège
Narvesen Info Center - NIC
Bertrand Narvesens vei 2
P.O. Box 6125 Etterstad
0602 Oslo 6   Tel. (02)57.33.00
Telex: 79668 NIC N   Telefax: (02)68.19.01

Pakistan
Mirza Book Agency
65 Shahrah Quaid-E-Azam
Lahore 3   Tel. 66839
Telex: 44886 UBL PK. Attn: MIRZA BK

Portugal
Livraria Portugal
Rua do Carmo 70-74
Apart. 2681
1117 Lisboa Codex   Tel.: 347.49.82/3/4/5
Telefax: (01) 347.02.64

Singapore/Malaysia – Singapour/Malaisie
See "Malaysia/Singapore" – Voir «Malaisie/Singapour»

Spain – Espagne
Mundi-Prensa Libros S.A.
Castelló 37, Apartado 1223
Madrid 28001   Tel. (91) 431.33.99
Telex: 49370 MPLI   Telefax: 575.39.98
Libreria Internacional AEDOS
Consejo de Ciento 391
08009-Barcelona   Tel. (93) 301.86.15
Telefax: (93) 317.01.41

Sri Lanka
Centre for Policy Research
c/o Mercantile Credit Ltd.
55, Janadhipathi Mawatha
Colombo 1   Tel. 438471-9, 440346
Telex: 21138 VAVALEX CE   Telefax: 94.1.448900

Sweden – Suède
Fritzes Fackboksföretaget
Box 16356
Regeringsgatan 12
103 27 Stockholm   Tel. (08)23.89.00
Telex: 12387   Telefax: (08)20.50.21
Subscription Agency/Abonnements:
Wennergren-Williams AB
Nordenflychtsvägen 74
Box 30004
104 25 Stockholm   Tel. (08)13.67.00
Telex: 19937   Telefax: (08)618.62.32

Switzerland – Suisse
OECD Publications and Information Centre
Schedestrasse 7
D-W 5300 Bonn 1 (Germany)   Tel. (49.228)21.60.45
Telefax: (49.228)26.11.04
Librairie Payot
6 rue Grenus
1211 Genève 11   Tel. (022)731.89.50
Telex: 28356
Subscription Agency - Service des Abonnements
Naville S.A.
7, rue Lévrier
1201 Genève   Tel.: (022) 732.24.00
Telefax: (022) 738.48.03
Maditec S.A.
Chemin des Palettes 4
1020 Renens/Lausanne   Tel. (021)635.08.65
Telefax: (021)635.07.80
United Nations Bookshop/Librairie des Nations-Unies
Palais des Nations
1211 Genève 10   Tel. (022)734.60.11 (ext. 48.72)
Telex: 289696 (Attn: Sales)   Telefax: (022)733.98,79

Taiwan – Formose
Good Faith Worldwide Int'l. Co. Ltd.
9th Floor, No. 118, Sec. 2
Chung Hsiao E. Road
Taipei   Tel. 391.7396/391.7397
Telefax: (02) 394.9176

Thailand – Thaïlande
Suksit Siam Co. Ltd.
1715 Rama IV Road, Samyan
Bangkok 5   Tel. 251.1630

Turkey – Turquie
Kültur Yayinlari Is-Türk Ltd. Sti.
Atatürk Bulvari No. 191/Kat. 21
Kavaklidere/Ankara   Tel. 25.07.60
Dolmabahce Cad. No. 29
Besiktas/Istanbul   Tel. 160.71,88
Telex: 43482B

United Kingdom – Royaume-Uni
HMSO
Gen. enquiries   Tel. (071) 873 0011
Postal orders only:
P.O. Box 276, London SW8 5DT
Personal Callers HMSO Bookshop
49 High Holborn, London WC1V 6HB
Telex: 297138   Telefax: 071 873 2000
Branches at: Belfast, Birmingham, Bristol, Edinburgh,
Manchester

United States – États-Unis
OECD Publications and Information Centre
2001 L Street N.W., Suite 700
Washington, D.C. 20036-4910   Tel. (202)785.6323
Telefax: (202)785.0350

Venezuela
Libreria del Este
Avda F. Miranda 52, Aptdo. 60337
Edificio Galipán
Caracas 106   Tel. 951.1705/951.2307/951.1297
Telegram: Libreste Caracas

Yugoslavia – Yougoslavie
Jugoslovenska Knjiga
Knez Mihajlova 2, P.O. Box 36
Beograd   Tel.: (011)621.992
Telex: 12466 jk bgd   Telefax: (011)625.970

Orders and inquiries from countries where Distributors
have not yet been appointed should be sent to: OECD
Publications Service, 2 rue André-Pascal, 75775 Paris
Cedex 16, France.

Les commandes provenant de pays où l'OCDE n'a pas
encore désigné de distributeur devraient être adressées à :
OCDE, Service des Publications, 2, rue André-Pascal,
75775 Paris Cédex 16, France.

75669-4/91

OECD PUBLICATIONS, 2 rue André-Pascal, 75775 PARIS CEDEX 16
PRINTED IN FRANCE
(66 91 07 3) ISBN 92-64-03349-1 - No. 45551 1991

300527260P